Praise

The Anatomy (

"*The Anatomy of Anxiety* is a paradigm shift in how we view anxiety and the mind-body connection. Dr. Vora's approach gets to often unaddressed root causes of anxiety and teaches you how to find well-being. Anxiety truly is our teacher, and this book is a must-have for anyone looking to heal."

—Dr. Nicole LePera, author of the #1 *New York Times* bestseller *How to Do the Work*

"The most comprehensive book in print on anxiety: what it is, its myriad causes, the ways in which it affects us, and how we can practically use, manage, and transcend it. *The Anatomy of Anxiety* is like *The Body Keeps the Score* for anxiety. Everyone needs this book."

—Holly Whitaker, *New York Times* bestselling author of *Quit Like a Woman*

"Dr. Ellen Vora is the rare psychiatrist who moves pas the prescription pad to probe the underlying factors that contribute to pervasive feelings of uneasiness. This is a hopeful guide to anxiety, one that addresses the ingredients of this complicated stew and offers tactical advice for unwinding its death grip on our minds and lives."

—Elise Loehnen, Former CCO of goop and host of the *Pulling the Thread* podcast

"At a time when anxiety is at an all-time high, my esteemed colleague Dr. Ellen Vora has shown impeccable timing and message with this book. This brilliant work offers us a holistic approach to anxiety that is thoughtful and comprehensive yet imbued with a beautiful grace and heart, just like its author."

—Dr. Will Cole, leading functional medicine expert and *New York Times* bestselling author of *Intuitive Fasting*

"What sweet salve—for anxiety to be seen as the 'beginning of inquiry,' not as a broken state. Thank you, Ellen, for leading us on this new, somatic, integrated, and vibrant journey with your expertise and your mindfully curated lived experience."

—Sarah Wilson, *New York Times* bestselling author of *First, We Make the Beast Beautiful*

"Anxiety keeps your life small. Dr. Vora's approach will get you out of your head and into your body, relationships, and lifestyle by dismantling anxiety (major and minor) for good."

—Melissa Hartwig Urban, CEO and cofounder of Whole30

"A thought-provoking debut. . . . Vora's sensible, easy-to-implement advice is backed up with solid medical rationales. Readers struggling with anxiety would do well to seek out this first-rate primer."

—*Publishers Weekly*

"Warm and highly readable. . . . Vora writes with compassion and is rarely prescriptive, noting that acting on even a few of her suggestions will be beneficial. . . . An open-minded and well-rounded approach to the psyche."

—*The Guardian*

The Anatomy *of* Anxiety

UNDERSTANDING AND OVERCOMING
THE BODY'S FEAR RESPONSE

Ellen Vora, MD

HARPER

NEW YORK • LONDON • TORONTO • SYDNEY

HARPER

A hardcover edition of this book was published in 2022 by Harper, an imprint of HarperCollins Publishers.

FIRST HARPER PAPERBACKS EDITION PUBLISHED 2025.

Designed by Bonni Leon-Berman

Library of Congress Cataloging-in-Publication Data has been applied for.

ISBN 978-0-06-307510-8 (pbk.)

25 26 27 28 29 LBC 6 5 4 3 2

For my mom

Contents

PART III: TRUE ANXIETY

The Anatomy *of* Anxiety

Introduction

We are on the cusp of a significant shift in how we view and treat mental health. Over the last several decades, the emerging fields of functional and integrative medicine, nutritional psychiatry, and even psychedelic therapy have shone a light on new paths to better mental health. These disciplines have demonstrated that issues we'd once considered purely psychiatric in nature can be better understood as the result of a delicate yet highly consequential interplay of body and mind.

In my own work as a holistic psychiatrist, for instance, I examine the whole portrait of my patients' lives—from what they eat; to how they sleep; to the quality of their relationships; to where they find meaning, purpose, and refuge in their lives. In doing so, I have found that the anxiety that plagues so many of us is increasingly caused by the habits that are now intrinsic to our modern lives, such as chronic sleep deprivation, poor nutrition, and even doom scrolling on social media late into the night. Though these issues may seem too benign to significantly affect the mind, they are capable of creating a stress response in the *body*, which prompts the release of hormones such as cortisol and adrenaline—signaling a state of emergency to the brain that can leave us feeling anxious. In other words: physical health *is* mental health. And anxiety—that hypervigilant feeling that escalates swiftly to a sense of catastrophe and doom—is as grounded in the body as it is in the mind.

This paradigm shift is, in my view, as revolutionary as when selective serotonin reuptake inhibitors (SSRIs), a class of anti-depressant medications including Prozac and Lexapro, were introduced a few decades ago. When these medications became the mainstream treatment for depression and anxiety, a clear medical model was presented for psychiatric disorders, and public mental health awareness grew. After centuries of stigma and shame, this came as an enormous relief; it offered the notion that our mental health struggles are not based on our personal failings but are essentially an expression of our brain chemistry. Now, however, given our increasing understanding of the profound mind-body connection, we have even more avenues to explore, in addition to medication, for addressing mental health. And in understanding that the body is as capable of informing our moods as the brain is, we have also come to realize that our anxiety is far more *preventable* than previously known. That is, through relatively straightforward adjustments to our diet and lifestyle, we can avoid unnecessary stress responses and head anxiety off at the pass.

There is, of course, a more profound anxiety that exists beyond the physiological—and this feeling of uncertainty and unease cannot be addressed quite so easily. I've found, however, that once I've worked with my patients to eliminate the first layer of physical anxiety, the way is cleared for us to tap into this more penetrating distress. When my patients are able to discern the message of this deep-seated anxiety, they often find that it is their inner wisdom sending up a flare that something is out of alignment in their lives, either with their relationships or jobs or in the world at large. It sometimes speaks to our estrangement from community or nature; at other times, it points to a lack of self-acceptance or a keen awareness of the grave injustices happening around us. Exploring this anxiety allows us to excavate our intimate truths. And more often than not, these revelations offer a call to action as well as an opportunity to turn a feeling of profound disquiet into something purposeful.

In this sense, whether it's the consequence of our habits or a missive from our inner psyche, anxiety is not the final diagnosis but rather the beginning of our inquiry. That is, anxiety is not what's wrong with you—it is your body and mind fiercely alerting you to the fact that *something else* is wrong. It is evidence that there is something out of balance in your body, mind, life, or surroundings—and with curiosity and experimentation, you can work toward putting these elements back into balance. The path forward begins with identifying the root cause, whether it is the result of an everyday habit or a sense of profound unease or both.

I come by these revelations honestly. My years in medical school at Columbia University and my psychiatry residency at Mount Sinai were not halcyon days for me, largely because my arduous training was complicated by my own mental and physical health problems. I struggled with mood as well as digestive, hormonal, and inflammatory issues—problems that I now know conventional medicine is fundamentally ill equipped to address.

It took me years to regain balance within my body and life. Ultimately, in the final year of my psychiatry residency, desperate to bring more meaning to my work as well as to discover a way to heal myself, I began studying alternative approaches to health in addition to my rotations at the hospital. When I wasn't working overnight on the wards, I attended acupuncture school and then took shifts administering acupuncture to patients at an addiction clinic in the Bronx; I used my elective time to complete the integrative medicine training at Andrew Weil's center at the University of Arizona and then had a mentorship with an integrative psychiatrist back in New York; I apprenticed with a hypnotherapist; I undertook intensive yoga teacher training in Bali, where I was also introduced to Ayurveda; and, in time, I went on to study functional medicine and explore psychedelic medicine and its potential implications for psychiatry.

If I had not fought to create this unique path for myself, I would not have learned about these other approaches to healing. In my nine years of medical school, research fellowship, and residency, not even one lecture was dedicated to discussing these modalities from other cultures and traditions. Yet when I was immersed in my alternative training, I felt as if I was expanding my medical perspective in a critical way. By taking on these practices in my own life, not only did I see a way forward to helping my patients thrive but I felt physically healthier than I'd ever felt in my adult life. The benefits I experienced seemed to outclass any improvements gained from an array of conventional interventions. And these learnings coalesced into the multifaceted, holistic approach to mental health that I offer in my practice—and that informs every page of this book.

Over the last decade, I have seen patients with varying circumstances and degrees of anxiety; most of them have been able to successfully address their mental health by first taking a look at their daily habits and then, if necessary, delving deeper into their emotional lives. There are those with whom I have worked only briefly, such as the twenty-five-year-old woman who came to me with a history of anxiety, digestive issues, and a mysterious rash. We sifted through her diet and identified and removed the inflammatory foods; within a month, her digestion came back online, her rash was gone, and her anxiety had abated. On the other end of the spectrum, I worked for several years with a woman I'll call Janelle who came to me in her mid-thirties, after having been involuntarily hospitalized for a manic episode; she was diagnosed with bipolar disorder and heavily medicated. Together, Janelle and I discovered that, in actuality, she had Hashimoto's thyroiditis, a condition in which your immune system attacks the thyroid and one that can present as alternating states of depression and activated anxiety—resembling bipolar disorder. We worked on changing her diet and lifestyle to heal her thyroid as well as slowly tapering her off her mood stabilizer. Janelle's anxiety has decreased notably, and she hasn't had a manic episode since. I also treat a young man who began therapy initially to explore his traumatic childhood; ultimately, however, we found our way toward exploring his gift of sensitivity, and he has since changed his career in order to help others work through trauma. By learning to differentiate between the anxiety that begins in the body and the anxiety that acts as a North Star, my patients are able to move forward into more expansive lives.

This book will offer practical and actionable steps to help mitigate anxiety. Given that the challenges of accessibility and affordability in mental health care persist, I make every effort to offer tools that are within reach. While it is always recommended to seek the support of a mental health professional for serious mental health issues, many of the fixes I propose here are inexpensive and can be done on your own (as well as with the support of a mental health professional, should you so choose). And yet, just because there are many things you *can* do, does not mean there are many things you *must* do. I have laid out what I have found to be the most effective and impactful interventions from my practice—but you should choose the strategies that feel right to you. What feels most approachable and suited to your needs? Feel free to skip a section if it feels overwhelming, and perhaps return to it later. Start with something that seems, if not easy, at least doable for now. With every change that you adopt, your anxiety will incrementally improve, making the next adjustment feel easier. In other words, I invite you to approach this book as you would a buffet: serve yourself what you feel drawn to, and you won't go wrong.

Most importantly, I encourage you to consider anxiety as an invitation to explore what might be subtly out of balance in your body and life. My hope is that this book will allow you to become more attuned to what your anxiety has been trying to tell you. I am not saying this will be simple—bodies and lives are complicated, and change can be hard. But there are now more opportunities than ever for ameliorating mental health issues, and I'm hopeful that among them there is a path forward for you to heal and be well.

PART I

It's *Not* All in Your Head

Chapter 1

The Age of Anxiety

Problems that remain persistently insoluble
should always be suspected as questions
asked in the wrong way.

—*Alan Watts*, The Book: On the Taboo against
Knowing Who You Are

We are in an unprecedented global crisis when it comes to mental health. An estimated one out of every nine people, or eight hundred million people, suffer from a mental health disorder, the most common of which is anxiety. Indeed, almost three hundred million people worldwide struggle with an anxiety disorder.[1] And the United States is one of the most anxious countries of all: up to 33.7 percent of Americans are affected by an anxiety disorder in their lifetime.[2] In fact, from 2008 to 2018, incidences of anxiety in the United States increased by 30 percent, including an incredible 84 percent jump among eighteen-to-twenty-five-year-olds.[3] Not to mention that the recent COVID-19 pandemic served to steeply escalate already dire circumstances. The number of people reporting symptoms of anxiety and depression skyrocketed by an extraordinary 270 percent, as researchers at the Kaiser Family Foundation found when comparing 2019 to 2021.[4]

And yet, while these statistics paint a grim picture, they also offer a reason to feel hopeful. These rates would not have risen so precipitously if these disorders had a predominantly genetic basis—which was our presiding understanding over the last several decades. Our genes cannot adapt so quickly as to account for our recent catapult into anxiety. It stands to reason then that we are increasingly anxious because of the new pressures and exposures of modern life—such as chronic stress, inflammation, and social isolation. So, odd as it may sound, this recent acceleration is actually *good* news because it means there are straightforward changes that we can make—from a shift in diet and sleep routines all the way down to better managing our relationship with our phones—in order to have a powerful impact on our collective mood. By widening the lens of our understanding to encompass not only the aspects of anxiety that occur in the brain but also those that originate *in the body*, we can more effectively address our current, and vast, mental health epidemic.

WHAT DO WE MEAN BY "ANXIETY"?

Anxiety has been recognized as far back as 45 BC, when the Roman philosopher Marcus Tullius Cicero wrote in the *Tusculan Disputations*, as translated from the Latin, "Affliction, worry and anxiety are called disorders, on account of the analogy between a troubled mind and a diseased body."[5] It is interesting that he would mention the body, given that anxiety has since then wended its way through history primarily understood as a problem of the mind; it is only now, twenty centuries later, that we are returning to the notion that the body plays a critical role in determining our mental health. The word *anxiety* derives from the Latin word *angor* and its verb *ango*, meaning "to constrict"; in fact, in the Bible, Job describes his anxiety correspondingly as "the narrowness of my spirit." As time wore on, the term *anxiety* became more closely affiliated with a sense of impending doom, or as Joseph Lévy-Valensi, a French historian of psychiatry, described it, "a dark and distressing feeling of expectation."[6] This definition remained largely unchanged throughout modern history, though it became increasingly clinical in its description once the disorder was introduced in the *DSM-1*, the *Diagnostic and Statistical Manual of Mental Disorders*, published in 1952. In the *DSM-5*, its most recent incarnation, *anxiety* is familiarly defined as "the anticipation of future threat," but the disorder is also broken up into an array of classifications, such as generalized anxiety disorder, social anxiety, panic disorder, obsessive-compulsive disorder (OCD), and post-traumatic stress disorder (PTSD).[7] Modern conventional psychiatry uses this multiplicity of classifications to steer treatment.

In my practice, however, I don't use such specific designations in naming my patients' anxieties. Though some believe that "anxiety" has become diluted or too all-embracing—indicating almost any feeling of discomfort—I think that the term *cannot* be too broadly used. If you're asking the question *Do I have clinical anxiety?*, I believe you are suffering in a meaningful way. I want you to trust your subjective experience of uneasiness more than I want you to worry about whether or not you qualify for a diagnosis. Over the years, I have seen anxiety expressed in such a myriad of ways among my patients that I have come to accept that it can be experienced as a vast and ever-shifting array of symptoms. I have patients who tell me that their lives feel generally fine—they are happy and healthy and have dynamic and supportive relationships—but they become paralyzed when under pressure at work. For them, anxiety—whether asserting itself as "impostor syndrome" or an inability to stop the mind from spinning in too many directions at once—serves as a barrier to dropping in and concentrating. I work with others who have anxiety solely around social life; some who *never* feel relaxed, constantly nagged as they are by some sort of dread or rumination; others who experience panic attacks out of the blue; and still others who feel only physical sensations—dizziness, light-headedness, or tightness in their chests or tension in their muscles. All these feelings are valid expressions of anxiety.

But there is also another critical reason I don't emphasize diagnosis in the work that I do. I have found that giving a diagnostic label—though it can offer immediate relief as a succinct interpretation of a fairly messy circumstance—can soon become a straitjacket of sorts, narrowly defining people and profoundly shaping their life narratives. Patients sometimes begin to conform their stories *toward* a diagnosis, making themselves smaller as opposed to opening to the more expansive lives they could be leading. So, ultimately, I'm less concerned with whether a person has *panic disorder with agoraphobia* or *OCD* or *generalized anxiety disorder* and more interested in exploring the particulars of each patient's life and habits in order to start them down a path to recovery.

TRUE/FALSE ANXIETY

And yet there *is* a distinction that I make within the realm of anxiety to help clarify what is being communicated to you by your body—and that is of *false* and *true* anxiety. This is not a diagnosis but rather an interpretation that I have found has helped my patients target the source of their unease and more swiftly identify the steps that need to be taken toward greater comfort and happiness. Julia Ross, a pioneer in nutritional therapy, first opened my eyes to this concept in her book *The Mood Cure*. Ross proposes that we have true emotions and false moods. True emotions occur when something steeply challenging has happened: a family member passes away, and you're grieving; you lose your job, and you're stressed; you're going through a breakup, and you're sad. These "genuine responses to the real difficulties we encounter in life can be hard to take," writes Ross, "but they can also be vitally important."[8] A false mood, on the other hand, is more like an "emotional impostor," as Ross puts it, when we seem to just wake up on the wrong side of the bed or, seemingly out of nowhere, find ourselves feeling irritable, sad, angry, or anxious about things that wouldn't normally trip us up. At these times, our minds are all too happy to swoop in with an explanation. Our brain says, *Maybe I'm anxious because my boss's aloof email seems to suggest I'm underperforming at work*; or *Something about that text from an old friend is not sitting right with me*. Our minds are meaning makers. Give us a picture of two dots and a line, and our minds see a face; give us a hangover and a cold brew in lieu of breakfast and we think we're in trouble at work, our relationship is falling apart, or the world is doomed, because our minds like to tell us stories that explain our physical sensations. And much of our worry is just this: our minds trying to justify a stress response in the physical body.

Ross's paradigm can also be applied exclusively to anxiety. False anxiety is the body communicating that there is a physiological imbalance, usually through a stress response, whereas true anxiety is the body communicating an essential message about our lives. In false anxiety, the stress response transmits signals up to our brain telling us, *Something is not right.* And our brain, in turn, offers a narrative for why we feel uneasy. It tells us we're anxious because of our work or our health or the state of the world. But the truth is, there is always *something* to feel uneasy about. And the reason we're struck with anxiety in *this moment* actually has nothing to do with the office and everything to do with a state of physiological imbalance in the body—something as simple as a blood sugar crash or a bout of gut inflammation. Much of our anxiety, in this sense, is unrelated to what we think it's about.

But let me make a critical clarification: just because I refer to these sensations as false anxiety *does not* mean the pain or suffering is any less real. Even if a mood is the direct result of a physiological stress response, it can still hurt like hell. This term is not meant to invalidate the experience of these moods. The reason I feel it's important to identify these states as false is that it allows us to see a clear and immediate path out. This type of anxiety is not here to tell you something meaningful about your deeper self; rather, it's offering a more fundamental message about your body. And when we recognize that we are experiencing anxiety precipitated by a physiological stress response, we can address the problem at the level of the body, by altering our diet or getting more sunshine or sleep. In other words, false anxiety is common, it causes immense suffering, and it's mostly avoidable.

Once we are able to target and eliminate this physiological source of our distress, we can then more directly address the deeper anxiety—our true anxiety—that arises from having strayed from a vital sense of purpose and meaning. At base, this anxiety is what it means to be human—to know the inherent vulnerability of walking this earth, that we can lose the people we love and that we, too, will one day die. Or as the nineteenth-century Danish existentialist philosopher Søren Kierkegaard described it, "the dizziness of freedom." This anxiety also, in some ways, keeps us safe. We are all here, after all, because our ancestors were vigilant enough to survive; this anxiety can fuel us to protect ourselves and to keep our lives in motion. But it also often arrives with a message—with intuition and wisdom from deep within—about what we need to do to bring our lives into more alignment with our particular abilities and purposes; it is essentially a guide for how to make our lives as full as they can be.

Chapter 2
Avoidable Anxiety

It isn't disrespectful to the complexity of
existence to suggest that despair is, at times,
just low blood sugar and exhaustion.

—*Alain de Botton*

When we are anxious, it can feel like everything is conspiring to overwhelm us: our relationships confound us, work presses and prods us, the world feels like it's barreling toward certain disaster. But many of the dreadful feelings and terrifying thoughts we call anxiety are simply the brain's interpretation of a fairly straightforward physiological process that comprises the stress response. And yet, in traditional psychiatry, doctors are trained to treat mental health problems by addressing solely the mind, with medications to alter brain chemistry and therapies to target thoughts and behaviors. As a result, most psychiatrists have implicitly learned not to overstep their bounds and get involved with matters of the physical body. I believe, however, that this approach has held the field back, limiting psychiatrists' treatment options when there is such an extensive range of ways to treat the mind *through* the body.

With the rise of integrative and functional medicine—and the newly burgeoning field of holistic psychiatry—we have begun to understand mental health disorders anew. Indeed, the evidence, not to mention the demand by patients, to take a more holistic approach to mental health has been mounting. For instance, a 2017 study known as the SMILES trial (an acronym for Supporting the Modification of lifestyle in Lowered Emotional States), led by Felice Jacka, an associate professor of nutrition and epidemiological psychiatry at Deakin University in Australia, looked at the impact of improving nutrition compared with social support in people with moderate to severe depression, all of whom ate diets of primarily processed foods; ultimately the researchers found that 32 percent of those receiving dietary support achieved remission compared with 8 percent in the social support group.[1] Similarly, in a number of different studies, the spice turmeric—used for centuries in Ayurvedic medicine, the ancient healing practices of the Indian subcontinent—was shown to have the ability to decrease inflammation and thereby modulate neurotransmitter concentrations involved in the pathophysiology of depression and anxiety.[2] (Inflammation occurs when the immune system is mobilized to address a threat, such as injury or infection, and it can directly signal that the body needs to fight back, leaving us feeling anxious.) So, while brain chemistry and thought patterns *do* play a role in anxiety, I would argue that these are often "downstream" effects—meaning that much of the time our brain chemistry changes *as a result* of an imbalance in the body. In other words, the root cause of false anxiety begins in the body, and it should be treated there as well.

THE SCIENCE OF FALSE ANXIETY

The general understanding in conventional psychiatry is that anxiety is largely the result of a genetic chemical imbalance in the brain. But there isn't consensus on the mechanisms *causing* anxiety, aside from the consistent focus on the neurotransmitter serotonin. However, there is another neurotransmitter, GABA (gamma-aminobutyric acid), that serves as the primary inhibitory chemical messenger of the central nervous system—and this, too, plays a critical role in assuaging our nerves. In my opinion, GABA hardly gets the attention it deserves, at least in our public discourse, given what a critical natural resource it is for battling anxiety. The effect of this neurotransmitter is to create a sense of calm and ease and, therefore, it has the power to inhibit an anxiety spiral. So, when we start dreaming up all the terrible worst-case scenarios that could occur in our lives, GABA can whisper to us, *Shhh, no need to worry, that's not likely; everything will be fine.* Conventional psychiatry, as a result, often deduces that a person experiencing anxiety has poor serotonin or GABA signaling and, ultimately, is not getting enough of the reassurances these neurotransmitters have to offer. It is my belief, however, that false anxiety is less about genetic destiny and more about the circumstances presented by our modern lifestyles—from taking a course of antibiotics to the chronic, unrelenting stress that so many of us are under. Not only do these assaults on the body diminish production of GABA—as I'll explore further in a moment—but there are also other pathways by which the body communicates to the brain that *things are not OK*. Two of the main physiological processes that provoke anxiety are the stress response—our nervous system's reaction to a perceived threat—and gut-related systemic inflammation.

THE STRESS RESPONSE

We typically think of the stress response as occurring as an automatic reaction to external events—such as bad news or a physical threat—but it can also be caused by various internal states of imbalance in the body, such as sleep deprivation[3] or even just a strong cup of coffee (which can prompt the body to release cortisol, the body's main stress hormone[4]). This revelation would seem anticlimactic if it were not such good news, because these physical causes of stress and anxiety are *preventable*. The stress response was hardwired into our bodies over millions of years of evolution to help us stay safe in the face of life-threatening situations, such as the presence of a menacing predator, which used to be a feature of daily life. This reaction begins with a hormonal cascade that is now familiarly known as the "fight-or-flight" response. The body anticipates needing immediately to engage in attack or, alternatively, to run. To allow for this, the body shunts blood flow away from places such as the gastrointestinal (GI) tract and genitalia and directs it instead toward the muscles, heart, lungs, eyes, and brain, making it possible to fight harder, run faster, see better, and generally outsmart whatever is posing the impending threat. The stress response achieves this by furiously pumping out hormones such as epinephrine—i.e., adrenaline—and norepinephrine, which dilate our pupils and the blood vessels in our muscles, while constricting blood vessels in our gut and skin, as well as cortisol, making us feel alert and mobilizing blood sugar to make energy available. Meanwhile, the amygdala, which is a subset of the limbic system—the part of the brain involved with processing emotions, memory, and behavior needed for survival—also goes to work, making our surroundings feel even more threatening.

Though today we have the same physiological equipment to cope with stress, we have a very different world to cope *with*. We encounter chronic, low-grade stressors rather than acute, life-or-death ones, such as inflammatory food, inadequate sleep, and an influx of messages coming at us through email, text, and Slack. And although these stressors are not as severe as facing a leopard, they still initiate the stress response. Whether the perceived danger is large or small, the body just keeps doing its job of preparing us to face a threat. Thus, with our modern diets and habits—which frequently trigger stress responses in our bodies—many of us live in a near-constant state of feeling under siege. Your blood sugar is crashing after eating something sweet? The body interprets this as a mild threat to survival. You stayed up too late doom scrolling on your phone? The body feels surrounded by danger. Sleep deprivation, chronic inflammation from eating foods you don't tolerate, and the comment section in Twitter—these are all, from your body's perspective, indications that your environment is not safe. So, the body releases stress hormones into your bloodstream, and this invisible chemical cascade manifests as the feelings and sensations of false anxiety.

Not only is this response largely avoidable but there are also ways to discharge the adrenaline that courses through our bodies after the stress response has been engaged, restoring us to a state of calm. Broadly, this is achieved by finishing the stress cycle, a concept that was recently popularized by Emily Nagoski, PhD, and Amelia Nagoski, DMA, sisters and coauthors of the book *Burnout: The Secret to Unlocking the Stress Cycle*, which posits that you must engage in an activity that tells the brain that, as the sisters Nagoski put it, "You have successfully survived the threat and now your body is a safe place to live"[5]—such activities include certain types of movement and self-expression. In

Part II, once we've learned more about how we can prevent a stress response—and its accompanying false anxiety—we will further explore particular techniques for finishing the cycle when stress is inevitable.

False Anxiety Inventory

As straightforward as the questions on this list may seem, my patients report that this inventory is one of the most impactful interventions for helping their anxiety. By pausing in the midst of turmoil and running through the following checklist of possible triggers, we can identify the particular false anxiety that might be occurring as well as understand its straightforward remedy. This process also helps take the charge out of feeling anxious, particularly when we can pinpoint a reason. I suggest patients keep this list on the fridge at home.

I'm anxious, and I'm not sure why. Am I . . .
- Hungry? (Eat something.)
- Sugar-crashing or having a chemical comedown? (Did I just eat something sweet, processed, or laden with food coloring or preservatives? Have a snack and focus on making different choices next time.)
- Overcaffeinated? (Perhaps this jittery anxiety is really caffeine sensitivity; tomorrow, drink less caffeine.)
- Undercaffeinated? (I drank less caffeine today than usual; dose up and aim for consistent daily caffeine consumption going forward.)
- Tired? (Take a nap; prioritize an earlier bedtime tonight.)
- Dehydrated? (Drink some water.)

- Feeling sluggish? (Take a quick walk outside; dance.)
- Dysregulated? (Did I just engage in an internet rabbit hole or social media binge? Dance or go outside to reset the nervous system.)
- Drunk or hungover? (File this away to help inform future choices around alcohol.)
- Due for a dose of psychiatric medication? (Right before the next dose, I'm at the pharmacological nadir—or the point where the level of medication in my bloodstream is at its lowest—and this can affect mood. Time to take meds.)

THE BRAIN, THE GUT, AND INFLAMMATION

As the last decade of scientific research has shown, the function of the gut and its microbiome, which is composed of trillions of microorganisms living in our intestines, goes well beyond simply digesting and absorbing our food. For one, the gut is the headquarters of our immune system, with more than 70 percent of our immune cells located in its wall.[6] The gut is also intimately connected to the endocrine system, including hormones that regulate our appetite, metabolism, and reproductive health. And, lastly, the gut is home to our enteric nervous system, increasingly referred to as "the second brain," which produces, uses, and modulates more than thirty neurotransmitters. In fact, this second brain creates and stores 95 percent of the serotonin in our bodies, whereas only 5 percent of our serotonin is found in the brain.[7]

Another critical aspect of gut health that is still largely underappreciated, however, is that the communication between the gut and the brain is a *two-way* street. Most of us understand the top-down communication; that is, when we're anxious, our digestion can be thrown off—consider the feeling of butterflies in your stomach when you're nervous or a bout of diarrhea before a big presentation. This occurs because our bodies have adapted to evacuating the bowels when facing a significant stressor, as this means there will be less weighing us down in a fight and less need for blood flow to the digestive tract, thereby allowing more blood to flow to the muscles, eyes, and heart. But just as the brain communicates to the gut, the gut also sends information *back up to the brain*. If the gut is calm and healthy, it sends an *all-clear* signal to the brain, allowing us to feel calm. If there's an imbalance among the microbes, however, or if we eat something our body doesn't tolerate, the message changes. In these instances, the gut can tell the brain: *feel anxious*.

This communication transpires largely via the vagus nerve—the longest cranial nerve in the body, running through the thorax and abdomen. In fact, about 80 percent of vagus nerve fibers are *afferent*, meaning that they gather information from the inner organs—such as the gut, liver, heart, and lungs—and deliver news about the state of affairs to the brain.[8] This essentially means that through the vagus nerve the gut has a direct hotline to the brain, letting it know what's happening at all times. If our gut is unhealthy, we'll feel uneasy.

This deeper understanding of the gut and its two-way communication with the brain helps us understand how dysbiosis, or an imbalance in gut flora—which is caused by such actions as taking a course of antibiotics, eating processed foods, or living under chronic stress—can directly contribute to our anxiety levels. There is even evidence that certain *Bacteroides* strains of gut bacteria—which are also compromised by poor diet and stress—are involved in the synthesis of the all-important neurotransmitter GABA;[9,10] in fact, because of our habits, I think of GABA as an endangered species of modern life.

But the gut has other avenues for sending an SOS to the brain in times of trouble. When the gut is irritated and inflamed, for instance, it can prompt the spread of inflammatory molecules, such as cytokines, throughout the body, causing systemic, widespread inflammation, which signals the brain to be anxious. One way this occurs is when endotoxins, officially named lipopolysaccharides, or LPS, pass through a compromised intestinal barrier, a condition popularly known as "leaky gut." Though endotoxins are normal inhabitants of a healthy gut, when they pass through the gut's barrier and reach the bloodstream—a state called endotoxemia—the immune system gets the alert that an invader is present, and it springs into action, inflaming the body and the brain.

Endotoxemia is not the only pathway through which the gut impacts our levels of inflammation and anxiety. The digestive tract also plays a central role in keeping the immune system calm, which is necessary to reduce inflammation in the brain.[11] A healthy and calm immune system depends on a diverse ecosystem of microbes in the gut. The population of beneficial bacteria, fungi, viruses, and even parasites[12] informs the immune system about tolerance and threat, teaching it when to relax and when to be concerned. In this way, the gut educates the immune system by teaching it to discern friend from foe. But if our gut is deficient in beneficial bacteria or overrun by pathogenic bacteria, our immune system misses out on this basic training and begins to misfire. An immune system run amok in this way can directly inflame our brain, as the inflammatory molecules travel to the brain via a network of vessels called the glymphatic system, sending up a direct flare that all is not right. Pronounced inflammation such as this makes us feel physically unwell through fatigue, aches, brain fog, or malaise—and also makes us anxious.

As such, diet and lifestyle factors are critical contributors to our mental health, given that they largely influence the state of the gut and our immune system. Our genes and our thoughts also, of course, have sway over our moods, but our daily habits are the real determinants of much of our anxiety. Ultimately, the more we can do to reduce stress in the body and inflammation in the gut, the better our chances of establishing a healthier mood. In Part II, we'll take a comprehensive tour of the strategies we can use to eliminate false anxiety, such as stabilizing our blood sugar to avoid unnecessary stress responses and healing our gut to decrease inflammation.

A WORD ON PSYCHIATRIC MEDICATIONS

Let me say this up front: I am grateful for the advent of antidepressants and other medications that treat mental health disorders. They provide much-needed relief for many people, some of my patients among them, and there are certainly circumstances in which medications are necessary and effective. Over ten years of practice, however, I have seen the efficacy of psychiatric medications play out in many ways, varying from patients who have been enormously helped by them to those who have experienced their potency fade over time to those who have not found them to have any effect at all to those who have suffered debilitating withdrawal effects. Given these wide-ranging scenarios, I am also grateful that our current understanding of anxiety—and the growing evidence that it is frequently based in the body—

has allowed me to adapt my practice to help many of my patients achieve a sense of balance and well-being through a shift in lifestyle.

Conventional psychiatry, however, has not quite caught up to this model and still treats anxiety as if it were predominantly the result of our thoughts and a genetically determined chemical imbalance in the brain—disregarding that much of anxiety is caused by physiological imbalances. But the truth is that psychiatric medications—targeted as they are to address a singular neurotransmitter, such as serotonin or GABA—*cannot* address false anxiety at its root cause. At best, they dampen the symptoms. I sometimes describe false anxiety as the *check engine* light of the body; rather than ignore this warning or cover it up with meds, it's generally preferable to fix the fundamental problem. I have found, too, that intervening at the level of the body, when indeed the source of a person's anxiety is physiological, is often quicker, less expensive, and more effective.

Later, in Part II, we'll take a closer look at psychiatric medications and their complexities, but here I would like to make clear that if you have been helped by psychiatric medications, there is no need to second-guess it. Count yourself as one of the lucky ones and take your meds. Alternatively, if you are planning on starting medication and you feel well informed about all the considerations, work with a caring practitioner and go forth. The strategies we'll cover can work alongside your meds, to address anxiety on two fronts. Importantly, when I discuss alternatives to medication, my intent is to help those who have not been helped by meds—not to make those who *are* helped question their choices. If you do fall into the category of someone who has not found meds to improve your mood, however, or you've experienced side effects, or you want to taper off for any reason, or you would simply like to try an alternate route, allow this book to guide you through the broad variety of options available for treating anxiety today.

YOU KNOW HOW TO HEAL YOURSELF

If an expanding array of treatment opportunities fills you with more anxiety, not less, let me put your mind at rest by assuring you that your body *wants* to heal. This book is also an effort toward helping people listen to what their bodies are telling them to do in order to recover equanimity and peace.

In a popular Instagram meme, a doctor perched on the side of his desk smugly says, "Don't mistake your Googling for my medical school education," to which the patient replies, "Don't mistake your one-hour lecture on my condition with my lifetime of living with it." Psychiatry can offer lifesaving interventions and useful support, but, frankly, the person who knows the most about your mental health is *you*. You are your own most powerful healer. This is a lot of responsibility, but it's also a relief.

I encourage you to trust in your knowledge of yourself as well as your body's wisdom and resiliency. What we experience as burdensome symptoms are so often signals of the body attempting to right itself, trying to get us back into homeostasis, the natural state of balance in the body. Rather than battle with our bodies, the goal should be to come to a place of mutual understanding and trust.

A patient once said to me, as we were discussing her multi-year struggle with an eating disorder, "I feel like I'm in couples counseling with my body." It struck me as an apt metaphor for my treatment philosophy. I hope my practice—and this book—

offers a kind of couples therapy between you and your body. It's a relationship in which, for many of us, communication and respect have broken down. There's resentment, frustration, mistrust, and rampant misunderstanding. So, let's take a page from couples counseling in working on this relationship: we need to listen to our bodies in order to understand what they need and what we can do to get back to a state of alignment. First, it's important to identify and address the physical anxiety—that is, the emotions we feel because the body's natural state has been overridden by unstable physiology, insufficient sleep, or nutritional neglect.

It is only once we've learned to avoid this unnecessary anxiety, after all, that we can tend to the deeper, true anxiety that remains. And though these emotions are less easily remedied, they also have more to offer by way of guidance, setting us on a path toward an abiding sense of purpose and fulfillment.

Chapter 3

Purposeful Anxiety

No matter how hard we try to ignore it, the
mind always knows truth and wants clarity.

—*Booker, in Toni Morrison's* God Help the Child

I have a patient, So-young, who came to see me to deal with her
anxiety, which was making it hard for her to focus at work, fall
asleep at night, and even enjoy her children. So-young grew up
in Queens, New York, the daughter of immigrant parents who
settled there after leaving South Korea. She told me that her
parents, who rebuilt their life in the United States from scratch,
placed a high value on outward appearances, and she felt that
their love for her was conditional, dependent on her looking and
behaving in a way that pleased them and impressed their community.

In her twenties, So-young married a difficult, narcissistic
man—someone who reflected back to her the familiar qualities
of her parents. As a result, throughout her marriage and while
raising two kids, So-young has battled near-constant anxiety.

When she first came to see me, she was already taking Paxil to level out her mood. I could sense that she channeled some of her anxiety through a need to put others at ease, as if it were her responsibility to make everyone in the room feel comfortable. I have to be a bit careful with patients like this—they are very easy to be around and can make my job feel effortless, but they also run the risk of not getting as much out of therapy as they need, given their people-pleasing tendencies.

We began to delve deeper into So-young's personal history as well as her relationship with her husband, exploring the roots of her unrest. So-young held a stance that is familiar to me from many of my patients: she believed she was simply genetically destined to be anxious. Her two sisters were also on medication, and her mother was prone to anxiety (though unwilling to engage with mental health care). One of So-young's sisters had told her, "We really have no choice in the matter—we will always need meds; it's just how we're wired."

While I believe without question that So-young's family has a genetic *predisposition* toward anxiety, I also believe that, as the saying goes, "genetics loads the gun and environment pulls the trigger." I suspected that So-young and I could identify and resolve the more fundamental reasons for her anxiety. After working together for a few months, So-young said that she wanted to taper off Paxil. This did not spring from a philosophical revelation about wanting to address her anxiety at the root; truth be told, when So-young learned that Paxil can cause weight gain, she wanted to be rid of it. Initially, as she slowly tapered off, So-young reported feeling a broader range of emotions, making her "feel more alive," and she was captivated by this early, unexpected shift. Over the next few months, So-young also began to describe her marriage with a stronger sense of indignation, identifying her husband's behavior as unacceptable and unlikely to change. Our sessions took a turn for the dramatic as she questioned whether to leave her husband and raise their children on her own. The people-pleasing aspect of her personality downshifted considerably. She wasn't reading the room as much as she was now drawing on her internal conviction and strength. Though I felt concern about the tumult in her marriage, I generally saw the changes in So-young as a positive sign of her wrestling with an authentic problem—primarily, that she'd been "asleep at the wheel," as she put it, often subduing her needs and making herself smaller in order to make room for her husband. As it dawned on So-young, however, that she had actually *needed* medication to tolerate her husband, she decided, with two young kids in the balance, to resume taking Paxil rather than face the tough road ahead.

Still, the seed had been sown. About a year later, in a passive bid to try coming off medication again, So-young didn't refill her Paxil when it ran out (not an approach to discontinuation I recommend, by the way). Again, after some time off her medication, she felt like she was waking up—but this time she fully committed to the process. "I feel like myself again," she would often say in our sessions, as if she was surprised and delighted to have stumbled on her "true self" once more. But she also saw the truth of her husband again—that he had an intense and demanding personality and felt entitled to having a compliant wife. This time, So-young chose to stand her ground and advocate for herself. She turned down plans with him and his friends in order to stay home and get some rest. She declined to have sex when she wasn't in the mood. She made plans with her own friends more often. "What has gotten into you?" her husband would ask. He pushed back against her new and surprising boundaries, and So-young explained what she was going through and the ways in which she hoped their relationship could change. Gradually, though not without a number of difficult conversations, her husband started treating her differently—better—and with more regard for her needs. By standing up for herself and recognizing that she was worthy of being considered with more respect, she successfully elicited the support she deserved from her husband. It has not been a straightforward path, but it has been life-affirming to see So-young more fully realized and, overall, shifting her marriage for the better. In her work with me, she is continuing to see her strengths and value beyond catering to everyone else's needs; she is realizing that other people's happiness is not her responsibility, and she's bolstering her belief in her own worth. Most importantly, she is less anxious, despite being off medication. I do not know if her marriage will continue to accommodate her as she changes and grows, but I do know—and more importantly, *she* knows—that she'll be able to guide herself with a steady and authentic sense of her needs as well as her implicit worthiness of having those needs met.

LISTENING TO ANXIETY

Even when we cut out the coffee and heal the gut, we will still be left with a certain amount of anxiety. This anxiety arises from the inherent fragility of life, but it also offers us the strength of our convictions. That is, when our lives don't align with our values or capabilities, we can feel anxious—but this feeling can also serve as a critical indicator that we need a course correction. Perhaps, like So-young, you are glossing over inequities in your partnership; maybe you are working in a job that fit your life when you were younger but now feels as if you took a wrong turn along the way; or you might feel unable to sit idly by as the planet continues to heat up and sea levels precipitously rise. Whatever the issue, this is your body's way of telling you, *Please look at this.* When you listen closely, this anxiety can point you in the direction of actions you need to take as well as the unique contribution you are here to make; ultimately, this feeling of unease can transform into a feeling of purpose. *This* is what I call true anxiety.

I tell my patients that they should embrace these feelings rather than try to suppress or avoid them. Instead of asking, *How can I stop feeling so anxious?*, we should be asking, *What is my anxiety telling me?* It is natural to reflexively resist this uncomfortable feeling. Culturally, we've also been taught to view anxiety as a nuisance, something to be suppressed into submission—but when we do this, we can miss out on critical guidance. What if you could learn to tolerate your anxiety long enough to hear what change is necessary? What if you could change the situation provoking your anxiety? What if instead of fearing and fighting true anxiety, you invite it in and hear what it has to say? Maybe you've been blocking something painful from your awareness, or maybe you just haven't slowed down enough to allow it to rise to the surface—but there is a part of you that has always known your essential truth. The "essential truth" of who we are has, as of late, become a bit of a cliché, spouted so often it can sound hollow. But for our purposes, I mean this as a buried instinct that, when too long ignored, can make itself known as mental discomfort—and that discomfort is trying to tell you something crucial.

The best way to hear the whisper of your intuition is by becoming still and quiet—it *will* eventually interrupt the nagging anxieties and chatter that play through your head on repeat. (In

Part III, we'll explore a variety of methods for connecting to this voice within yourself.) As you become familiar with this more resonant anxiety, you will come to feel it in your body too. When you experience warmth or a sense of expansion, that is often your body's way of saying "yes," of nodding in agreement with your gut feeling. When your body contracts, feeling tight or uneasy, that can be your true anxiety's way of tapping on your shoulder to indicate it still has not been fully heard.

True anxiety and intuition also generally register as a more substantial feeling. "My anxiety is high, it's like a shaky hovering, it's a high frequency . . . it's buzzing," the *New York Times* bestselling author and activist Glennon Doyle once said, describing the difference between her own fear and intuition. "But . . . there is something below it that is heavier, that is more grounded, that is not shaking, that is solid, that is the Knowing. And I actually now am at a time in my life—at 45 years old—where I can tell the difference."[1] In other words, even as true anxiety and intuition might be communicating to you that something is not right, they *feel* different from false anxiety. Instead of feeling like a threat, they come from a place of clarity and compassion.

If you choose to listen to true anxiety and let it steer you, it can be a golden compass, helping you navigate the vagaries of life. It allows for more growth, learning, and love. Transforming your true anxiety into something more purposeful does not, however, mean that things will necessarily get easier. For many of my patients, just as things start to get easier, they up-level to a more advanced set of challenges. They arrive at yet another growth stage, as happened to So-young in her marriage, where they feel out of place in familiar surroundings. Often, as you become more adept at using your true anxiety as a guide, life becomes more demanding because you are achieving more. This can sometimes feel excruciating. "It's like peeling away defenses that helped alleviate one layer of anxiety," my patient Ethan once said, "and I'm losing weapons as I go to face the monster." The *monster* in Ethan's case was childhood trauma, which he was ultimately able to face and release. Trauma, in particular, which we will explore in depth in Part III, occupies an unusual place in the true and false anxiety paradigm in that it exists at the intersection of the two. That is, traumatic experiences are often stored in the body—as the psychiatrist and bestselling author Bessel van der Kolk, MD, wrote in his groundbreaking book *The Body Keeps the Score*—which then also reprograms the brain. When this occurs, the amygdala—that part of the limbic system responsible for our fear response—is left in a state of hyperarousal, creating disproportionate anxiety throughout life. Trauma—which can occur from a range of experiences, from sexual assault to combat to emotional deprivation from a parent—leaves the brain on high alert, even if the threat is no longer present. As such, it has a false-anxiety aspect in that the brain can infer danger where there is none. And yet trauma should be treated as true anxiety, as the changes in the body were an adaptation to an unsafe world, and the hypervigilant amygdala is asking that the person reconnect with the trauma in order to arrive at a place of relative resolution. The feeling of true anxiety, as is the case with trauma, almost always has a larger historical context; that is, one episode of anxiety can hold decades of past life experiences within it, sometimes even longer. I have, indeed, worked with many patients who are unraveling the traumas of past generations that continue to leave an imprint on their lives—as well as unearthing the true anxiety still reverberating from the past. The truth can be a lot to hold; it can be difficult and destabilizing. This is our burden, as humans who dare to feel it all. But it is also how we expand as individuals, get in alignment with our purpose, and show each other the way forward.

TRUE ANXIETY IS YOUR SUPERPOWER

Studies of primates show that some members of the tribe are more anxious than others—these are the ones that tend to hang back, gathering at the peripheries of the main group. In the 1980s, the late zoologist Dian Fossey decided to remove these more sensitive members of one group of chimpanzees to see how it would affect the rest of the community. Six months later, all the chimps were dead. "It was suggested that the anxious chimps were pivotal for survival," Sarah Wilson writes compellingly of this experiment in her book *First, We Make the Beast Beautiful.* "Outsiders, they were the ones who were sleeping in the trees on the edge, on the border, on the boundary of the community. Hyper-sensitive and vigilant, the smallest noise freaked them out and disturbed them, so they were awake much of the night anyway. We label such symptoms anxiety, but back when we were in trees, they were the early warning system for the troop. They were the first to scream, 'Look out! Look out!'"[2]

Similarly, if you are one of the more attuned, anxious members of the human race—if your nervous system is dialed a little higher than others—the tribe owes you support and gratitude, because in important ways, your anxiety exists to protect us all. Instead of telling the anxious among us to "stop being so sensitive," we should honor what they have to say. The more every one of us embraces our true anxiety, the more valuable we are to our world. True anxiety does not just guide us on our own path, it assigns us a larger mission. Our true anxiety can place us on the front lines, alerting others to threats that may be just out of view. And the collective voice of true anxiety shepherds us in the right direction as a society.

And, objectively, the world needs to change. We are in the midst of a necessary reckoning. We have seen the Me Too movement pull back the curtain on sexual harassment and assault; the Black Lives Matter movement has opened up new and long-overdue dialogues about centuries of injustice and harm; and climate-change activists are shouting from the rooftops in an effort to be heard before it's too late. It is critical that we shift from pathologizing and suppressing this anxiety to heeding its urgent messages. We need to listen to those with their ears to the ground, who sense the subtle—and not so subtle—dangers on the horizon. They are our prophets, and they may just wake us all up in time.

IT'S A BOTH/AND

F. Scott Fitzgerald famously observed in his 1936 essay "The Crack-Up," "The test of a first-rate intelligence is the ability to hold two opposed ideas in the mind at the same time, and still retain the ability to function."[3] The same goes for anxiety: it's a both/and proposition in that it is possible to experience competing, and seemingly contradictory, states of anxiety—false and true—at the same time. Anxiety is physical. It's serotonin, GABA, gut inflammation, cortisol, and an overactive amygdala. But anxiety is also psychospiritual, existing at the interface between our psychology and our spiritual needs. It is about disconnection from purpose, from each other, and from ourselves. And no amount of gut healing, decaf coffee, or Paxil will touch these feelings. The only way to resolve this anxiety is to listen. It is OK—in fact, it is optimal—to identify and engage with both forms of anxiety simultaneously. There is no need to view your moods in just one way, and there is almost never just one right path. What I hope is that learning how to discern which type of anxiety is at play in any given moment, and how to respond accordingly, will help you know when to address your anxiety at the root—and when to slow down and listen to its urgent message.

PART II

False Anxiety

Chapter 4

The Anxiety of Modern Life

If they give you ruled paper,
write the other way.

—*Juan Ramón Jiménez*

As you now know, anxiety is a physiological phenomenon that involves both mind and body. In the body, anxiety manifests as a by-product of our stress response and the cascade of chemicals that are released when we confront a real or perceived threat. Anxiety can also be a result of other states of physiological imbalance—everything from inflammation and micronutrient deficiencies to hormone imbalance and impaired GABA transmission.

These physical forms of anxiety might be the most constant and burdensome, but they are also the most easily prevented and treated. Indeed, some anxiety is entirely *avoidable*. As such, the chapters that follow feature actionable strategies that can be implemented to not only alleviate false-anxiety symptoms but also bypass them altogether. I think of these interventions as the low-hanging fruit—the suggestions I offer patients at the beginning of treatment so that they can experience some quick wins and start to gain clearer headspace to tackle the more challenging true anxiety. We'll be discussing changes you can make to your sleep habits; your relationship to technology; the foods you eat; and the state of your gut, immune system, and hormones. Ultimately, we'll end with effective techniques for releasing the stress that we inevitably accumulate in our lives. Some of my patients don't even need to go further than these steps; their lives are in good order, but they need to modify their habits to better support their mental and physical health. These are the ones dealing exclusively with false anxiety, and they're able to feel better relatively quickly.

That said, even if you are "only" dealing with false anxiety, it's important to note that any change can be difficult to sustain—it is always easy to fall back into familiar routines. In particular, my recommendations about what to eat and drink present a significant challenge to many of my patients—but these changes are also often most impactful in abating their anxiety. But you don't need to heed every piece of advice I offer, nor do you need to make all the changes at once or in any particular order. Start with what feels most applicable to you and make a plan that best fits your circumstances.

The right amount of effort will be different for everyone. For example, if aside from experiencing anxiety, you're feeling physically well, and your body generally functions normally, then I don't think it's necessary to get your PhD in meal prep or make radical changes to your diet. On the other hand, if bodily dysfunction is standing in the way of living your best life, then it's worth quite a bit of effort to get your body back into balance.

Any small sacrifice you make in terms of diet will be more than repaid by the fact that you can now live without digestive issues and get through the day without panic. The ability to experience sustained comfort in your body is worth just about any initial inconvenience.

Most of my patients land somewhere in between these two poles—they aren't functioning optimally but aren't in an acute state of distress either. Perhaps this is where you are too. Before we proceed, it's worth taking a moment to check in with yourself. How much are you suffering? You should be aiming for effort and sacrifice that are proportional to your degree of discomfort. This is a deeply personal equation, but overall, our goal is to strike a quality-of-life balance.

It's also important to note that *ease* is itself a powerful agent of healing. Agonizing over every food choice certainly doesn't help to resolve anxiety. In the modern American food environment, I find this to be a very tricky line to toe. I want my patients to carefully consider their food choices and seek out the best-quality food they can find and afford, but I don't want them to be stressed out about every meal or worry about what will happen if they eat fast food while traveling. I tell them what I will tell you: do the best you can, and don't strive for perfection. We live in a country that makes it very difficult to eat well— our systems have failed us, and that can sometimes make it an impossible task to find healthy food, especially when you're not at home. Make the best choices you can with the circumstances you have—and be sure to allow yourself the occasional indulgence. If an ice cream cone or homemade cookies bring you joy, sometimes that matters more than avoiding a sugar crash.

We all have to find our own sense of stability. Perfect health is not in and of itself the goal; the goal is to feel good and have a fulfilling life. If your health is getting in the way of a fulfilling life, then let's roll up our sleeves and get to work. If the work itself of getting your body healthy is getting in the way of ful-fillment, then it's time to loosen your grip. With that balance in mind, let's take a look at the aspects of life that might be causing avoidable anxiety for you.

Chapter 5

Tired and Wired

I've started reading before bed instead
of scrolling Twitter and not only am
I sleeping really well, but I also think
I'm better than everyone.

—*Alex @alexgmurd*

We all know that getting a good night's sleep is essential to our well-being, but few of us realize how crucial this is for brain health. The connection between sleep and anxiety is a critical two-way conversation: anxiety contributes to insomnia, and chronic lack of sleep makes us prone to anxiety. Nearly forty million Americans struggle with chronic sleeplessness.[1] But with a few notable exceptions—shift labor, certain sleep disorders, and jet lag among them—modern insomnia woes are eminently fixable. Which is a good thing because there is probably no more effective or accessible treatment for anxiety than sleep. It's free, it feels good, and it *works*.

While there are a few folks out there who still haven't gotten the memo that they must prioritize sleep, what I see more often with my patients is that sleep eludes them despite their best intentions. For many of my patients, the challenge is that once they've added up work, commuting, cooking, childcare, life logistics, and perhaps a minute for decompression, there are simply not eight hours left for sleep. Even when the stars align and they manage to get to bed on time, they may lie there with thoughts racing or wake up in the middle of the night, unable to fall back asleep. The reality is that much of our insomnia is caused by our environment and the seemingly small choices we make every day. Fortunately, our body *wants* to sleep and knows how to do it—the trick is to listen to its cues, provide it with the right conditions, and get out of its way.

WHY SLEEP MATTERS

From a survival standpoint, sleep seems like a pretty maladaptive thing to do. Why on earth would we want to make ourselves vulnerable—prone and unconscious for eight straight hours in the dark—while our predators circle around? The fact that we do it, in spite of how dangerous it was in the early days of human evolution, points to its necessity—there *must* be a good reason our bodies need to restore in this way routinely, or else we wouldn't make ourselves so defenseless. Indeed, it's not possible to forgo sleep: sleep deprivation and disruptions cause severe cognitive and emotional problems, and animals denied sleep for several weeks ultimately die of infections and tissue lesions.[2] Though the scientific world hasn't yet entirely unlocked the black box of sleep's mysteries, it has uncovered a few of sleep's critical processes. We know, for instance, that memory consolidation occurs during sleep—by which we integrate the day's learnings into preexisting networks in the brain, putting them in long-term storage—and that the body repairs cells,[3] fights infections,[4] and restores energy during this time.[5,6]

But one of the most vital functions that sleep serves is to allow for the detoxification of the brain. At the end of a long day, your brain has worked hard. Thinking requires a lot of energy, and all that activity generates metabolic waste and toxic by-products, including beta-amyloid and tau oligomers—which are, incidentally, the same deposits found in the brains of Alzheimer's patients, albeit to a more severe degree. These accumulate in the brain during the day, and our glymphatic system, a subset of the lymphatic system found in the central nervous system, clears them away at night.[7,8,9] That is, as long as we're asleep. But if our sleep is compromised, our bodies never have a chance to clean up the mess created by our daily neurological heavy lifting.

Think of the brain as a little city. The activity in the homes and stores of the brain generates trash, and at the end of the day this trash is piled up in the alleyways. Then, while you sleep, garbage trucks—otherwise known as the glymphatic system—pick up and clear away the bags. If you don't sleep, though, the trash doesn't get picked up. So, the next day, you go about your life with garbage bags piled up in the alleyways of your brain. Your head vaguely hurts, it's hard to think clearly, you're less physically coordinated, and you're moody and anxious. If you've ever cared for a newborn baby, pulled an all-nighter cramming for a test, or worked the night shift, you know exactly what this feels like.

One commonly cited reason for not being able to sleep is stress—which is noteworthy because the neurotransmitter norepinephrine, a central player in our stress response, also plays a role in regulating the activity of the glymphatic system.[10] This implies that lack of sleep *and* chronic stress both can interrupt this necessary trash removal from the brain, potentially impairing the brain's ability to detoxify itself.[11] By managing stress *and* prioritizing sleep, we allow the glymphatic system to detoxify our brains on a regular basis—decreasing our baseline level of anxiety the next morning and potentially protecting ourselves from cognitive decline in the long term.

LET THERE BE DARKNESS

In the "boardroom of evolution," they sat around the table debating, *How are we gonna make the humans feel awake during the day and tired at night?* One member of the committee must have suggested using light as the cue—feel awake when the sun is shining and tired when it's dark. This suggestion was an instant hit and quickly became built into the human design.

It was a brilliant plan. And it was foolproof—until the advent of electricity. Then we got the lightbulb, the iPhone, and Netflix, and now nobody can sleep. The main reason we have difficulty sleeping is that our circadian rhythm, the body's internal clock, is cued by light, and in modern life it's getting all the wrong signals. We sit indoors in artificially lit spaces by day—hardly glimpsing the sun—then by night we find ourselves amid a psychedelic light show of TVs, laptops, and phones—all against a backdrop of ambient light pollution just outside our windows.

This is the process by which our body dictates its internal clock: there's a line of communication between the pineal gland—which secretes the sleep hormone melatonin—and a part of the brain called the suprachiasmatic nucleus (SCN), which is, in turn, directly connected to the eyes via the optic nerve. You can think of the SCN as our internal clock, and the primary method it has for telling time is to scan the landscape for light cues. Only when our eyes get a consistent input of darkness does the SCN become convinced that it's nighttime, prompting the *all-clear* signal for the pineal gland to secrete melatonin.

Think of melatonin as money—it's hard to earn and easy to spend. And, essentially, you earn melatonin by being enveloped in authentic, uninterrupted darkness in the evening, which is exactly what most of us are missing in modern life. With one glance at the phone or one flick of the bathroom light switch at night, your SCN tells your pineal gland, *Never mind, it's not actually nighttime*—and all your hard-earned melatonin is squandered. To get the circadian rhythm back on track and sleep well in modern life, we need to fix our melatonin by fixing our light cues.

We can do this by exposing our eyes to bright natural light first thing in the morning; this starts the internal clock, which cues the hormonal cascade that helps us feel awake during the day and tired at night. So, open your blinds as soon as you wake up. Spend some time outdoors in broad daylight and ditch your sunglasses. Ideally, get outside before 9:00 a.m., even if it's just for a quick walk around the block or two minutes on your front stoop.

At night, it's more of a challenge to re-create the conditions of early human life, but technology has its advantages as well as its drawbacks. Install dimmers and bring down the lights in your home after sunset. Program your devices with night-shift mode, and install a program such as f.lux on your computer to make the screen amber-hued at night (and thus less disruptive to your circadian rhythm). I realize that it's not easy or realistic to live in a blue light–free environment every evening; as I'm typing this paragraph on my laptop, it's 10:57 p.m. (I'm a mom with a full-time job and no childcare during a pandemic, so against all my own advice, this is when I write.) But I am also wearing blue light–blocking glasses, which have specially crafted lenses that filter the blue light from my computer and prevent it from impacting my circadian rhythm. I recommend anyone struggling with sleep to consider wearing these from sunset to bedtime.

This blue light is also the primary reason that I urge all my patients to *keep their phones outside their bedrooms.* Our ubiquitous phones emit blue light, thereby suppressing melatonin at night and disrupting our circadian rhythm.[12] Further, our endless scrolling before bed keeps us awake past the point when we're perfectly tired. We push past the just-right point of sleepiness into a state of being "overtired," where our body secretes cortisol, making it more difficult to fall asleep and stay asleep. And when the phone is on our bedside table, we glance at it anytime we wake in the night, sending a shock of blue-spectrum light to our brain like a shot of espresso. Setting up your charger outside the bedroom and committing to a phone-free bed seems intimidating in anticipation. Simply give this advice a one-week trial, and see whether you really miss your phone—and if you sleep better. If you rely on your phone as your alarm clock, experiment with using a good old-fashioned alarm clock instead.

Finally, when it comes time to turn off all the lights for bedtime, try to create as dark an environment as you can. Consider using an eye mask or blackout shades. Remove any unnecessary electronics from the bedroom. If you wake up in the middle of the night, try not to let your eyes "see" any light. If you have to go to the bathroom at 3:00 a.m., try the squinty shuffle—keeping your eyes open just enough to feel your way to the bathroom. If you need a nightlight, get an orange one.

HITTING RESET IN NATURE

My patient Travis struggled with severe insomnia for years. He tried *everything*. After we exhausted every lifestyle strategy, including eliminating caffeine and adding blue light–blocking glasses, and tried CBT-I (cognitive behavioral therapy for insomnia—an effective but intense strategy for addressing insomnia based on sleep restriction and sleep efficiency), we were still both left scratching our heads. It seemed like it was time to write this guy a prescription for Ambien. And yet I knew on some intuitive level that Travis's insomnia was related to the modern environment and, therefore, should be treatable through shifting his lifestyle—but we just couldn't make any headway. He worked as a software engineer, spending his days staring at a computer screen, and he lived in a high-rise NYC apartment, with ample light pollution from streetlamps and office buildings creeping in around his blackout shades. And then one day the idea came to me to suggest he go camping over a three-day weekend.

Travis pushed back immediately, explaining that he thought my recommendation seemed pretty extreme. If you ever have a psychiatrist tell you, from their comfortable recliner chair in their climate-controlled office, that you should go camping, I'll be the first to admit, that *is* a bit extreme.

But here's the thing: we can't argue with our genes. Since our bodies evolved with a circadian rhythm cued by contextual signals such as the sun and brightness by day and the moon and darkness at night—and modern life has turned the whole script upside down—it's always useful to find paths back to a more natural state. Camping restores those original contextual cues.

Camping is a lot like organic food. Which is to say that what was once just the way things were is now a pretty elaborate lifestyle *choice*. Organic food used to just be *what food was*—food grown in healthy soil without the use of chemicals; camping used to just be *what life was*—living in proximity to nature, sleeping close to the ground, with the abundance of sunshine during the day and darkness at night. So, from the perspective of our genes, *modern life* feels extreme, and sleeping under the stars feels like home.

In the end, Travis reluctantly went camping. And with the complete absence of artificial light after dark, he slept like a baby. That camping trip was six years ago. I check in on him from time to time, and he continues to sleep well—a new and much-appreciated rhythm for him that is supported by the occasional weekend camping trip. So, if you're struggling with insomnia and you feel like you've tried *everything*, pack your bag and take yourself into the wild. There is no more powerful way to set straight a confused circadian rhythm than to give it our original cues.

Shift Labor

For five years I worked overnight shifts at the hospital. I remember the culture of going out for brunch the next morning, soothing our frayed nerves with coffee. While coffee and pancakes may have provided a temporary high, we just felt even more fried and anxious in the long run.

Working overnight takes a considerable toll on your body, and in fact, people who work the night shift are more likely to suffer from a range of health issues including obesity,[13] cardiovascular disease,[14] and breast cancer.[15,16] One theory is that the light exposure overnight suppresses melatonin, which compromises our immune response and our body's ability to address nascent cancers. Being awake through the night also tips the balance of hormones such as leptin and ghrelin, involved in appetite, satiety, and metabolic health.[17] All of this is to say, the risks of working the night shift are not to be taken lightly. If you are required to work overnight shifts, my advice is to use your daylight hours to give your body what it needs to recover. To restore your basic physiology, it's important to develop some discipline around daytime sleep. When you leave work, wear blue light–blocking glasses, which will filter out the sun's blue rays that tell your brain it's daytime. After work, go straight home, draw your blackout shades, and get into bed wearing an eye mask. Do everything you can to convince your brain that it's nighttime, and give yourself the gift of restorative rest before you begin your next shift.

A CASE FOR AN EARLIER BEDTIME

In college I typically slept from 2:00 a.m. to 10:00 a.m., thinking, *As long as I'm getting my eight hours, it doesn't matter what schedule I'm on.* Never mind the fact that my body felt like a machine with the springs popping out of it. It turns out, the human body functions best on a schedule that syncs up with the sun.

When anthropologists study the last remaining preindustrial hunter-gatherer societies on earth, they consistently find something interesting: these tribespeople tend to go to sleep around three hours after sunset.[18] (They also, incidentally, do not have an equivalent word in their languages for "insomnia," as it is such a rare occurrence.) These populations are scattered across the globe, and they're definitely not tweeting about chronobiology, so they have independently arrived at this bedtime, making something about going to bed three hours after sunset seem inherently optimal for humans. Note that this is not a set time—it will vary by the time of year and where you are on the globe. Your bedtime might be 11:00 p.m. in June and 8:30 p.m. in December. If you prefer more of a one-size-fits-all recommendation, I suggest a bedtime around 9:30 p.m. to 10:45 p.m. for most people most of the year.

When we miss that window—whether it's because we're trying to get a few more things done, we're out having fun, or we're finding Netflix too tempting—our body goes into a stress response. It thinks, *There must be a good reason I'm not heading to bed when I'm tired; perhaps I'm in danger, or on the night shift watching over the tribe.* And the body releases cortisol, providing a jolt of energy and alertness to support our seeming need for wakefulness in what it thinks are exceptional circumstances. This is what it means to be "overtired." If you don't have kids, you may think to yourself, *Overtired . . . That's not a thing. You get tired, and then you get* more *tired.* The parents out there, on the other hand, know exactly what I'm talking about. When I had my daughter, I learned (the hard way) about this charming state when kids get so tired, they can't sleep. You see, when babies are tired, they have these cute little tired signs—they yawn and rub their eyes. But, actually, the situation isn't cute at all; it's practically an emergency. If you see a tired baby, you must drop whatever you're doing and get it in the crib before it's too late. Why? Because babies that go to sleep when they're perfectly tired will sleep. However, if you miss that window, babies become overtired—and try what you may, they will not sleep. And parents may not achieve basic goals of hygiene or adulting that day.

When babies get overtired, their bodies release cortisol, making them tired and wired. And, it turns out, grown-up bodies do the same thing. We are, in many ways, just giant toddlers after all. Do yourself a favor and acquaint yourself with your own tired signs. Does falling asleep on the couch sound familiar? For me, I've come to notice that I rub my eyebrows when I'm tired. When I push past that and get overtired, I get a second wind. I start to feel warm and suddenly find myself diving into the abyss of the internet or organizing the kitchen. When I finally do try to fall asleep, I toss and turn. It's as though I can feel my exhausted body battling with the cortisol coursing through my veins.

So, when you notice your tired signs, drop what you're doing and crawl into bed before it's too late. If you prefer a simple directive, here it is: Go to bed earlier.

Me Time, Not Screen Time

If you feel that the last forty-five minutes of putzing around on your phone at night is your only "me time" or opportunity to decompress, you're not alone. This is a habit among many of my patients of all ages and circumstances. In fact, there is a Chinese term for this behavior—報復性熬夜, which roughly translates as "revenge bedtime procrastination"—that occurs when "people who don't have much control over their daytime life refuse to sleep early in order to regain some sense of freedom during late night hours."[19] This astute concept offers its own solution in that it makes clear how important it is to find a pocket of time for yourself elsewhere in your day—one that isn't destructive to your sleep. How can we create daytime lives that we don't need to rebel against in the evening? Short of turning our capitalist society on its head, some of the solutions I've come up with for my patients range from carving out a few minutes in the morning for quiet reflection, to saying *no* to some social plans, to taking a twenty-minute walk after dinner, decompressing from the day.

NUTRITION AND SLEEP: HANGRY OVERNIGHT

So many of my patients tell me that even when they are able to fall asleep, they struggle to *stay* asleep through the night. There are many reasons why you may wake up in the middle of the night, but more often than not, sleep disruptions can be attributed to fluctuations in blood sugar.

The longest period of time between meals, for most of us, is when we sleep. While this fasted state is important for cellular repair and giving the digestive tract a rest, our blood sugar fluctuates overnight just as it does during the day. If you typically get "hangry"—angry and irritable when you're hungry—at 3:00 p.m., the overnight equivalent is waking up at 3:00 a.m. with racing thoughts, unable to fall back asleep. This typically happens when your blood sugar crashes overnight and your body counters with a stress response. In our four stages of sleep—which includes the three progressively deeper stages of NREM (non–rapid eye movement) sleep, and REM sleep, when we dream—a stress response can make your sleep more superficial, shunting you out of the deeper stages of sleep and making it more likely you'll be jolted awake. The solution is to maintain stable blood sugar throughout the night. How can you do this, short of feasting on a midnight snack? Well, generally, eating a blood sugar–stabilizing diet or retraining your physiology with intermittent fasting can prevent this issue before it happens. But in a pinch, my trick is to keep a jar of almond butter or coconut oil next to your bed and take a spoonful just before you brush your teeth at night. If you wake up in the middle of the night feeling jittery and anxious, take another spoonful. This bolus of fat and protein is slow to be digested and absorbed, and thus provides a steady safety net of blood sugar to carry you through the night without a blood sugar crash–induced stress response.

Espresso Nightcap

Many people are trapped in the vicious, addictive cycle of caffeine: exhausted every morning, we chug a cup; fading in the afternoon, we have another . . . only to find it's difficult to fall asleep at night. Bleary-eyed the next morning, we begin the cycle again. I'm no stranger to this ritual or to that feeling when your brain seems like a car that won't start, and coffee looks and smells like your only salvation. This feeling, however, is likely caffeine withdrawal. We have physiologically conditioned our bodies to *need* caffeine first thing in the morning, so they refuse to function without it.

You may be skeptical that one innocent cup of coffee in the morning could impact your sleep some fifteen hours later. But most people underestimate how long caffeine lingers in the body. Do you remember the concept of half-life from high school chemistry? Caffeine has an average half-life of about five hours,[20] meaning it took your body about five hours to metabolize *half* of the caffeine you drank this morning, and it will take another five to metabolize the next half, and so forth. This matters because it means some of your 9:00 a.m. latte is still buzzing around your brain overnight, and, more pressingly, drinking a cup of coffee at 3:30 p.m. is equivalent to drinking a half a cup at 8:30 p.m. Most people who struggle with insomnia wouldn't dare consume caffeine in the evening—and yet their daytime consumption has the same impact. Since even a little bit of caffeine can disrupt sleep, it's best to restrict your coffee habit to the early morning and decrease overall caffeine intake. Whether you're having a seemingly benign cup of coffee in the morning, feeling virtuous drinking matcha (which is basically Instagrammable rocket fuel) in the afternoon, or sipping Diet Coke at night, it's all consequential for your sleep *and* your anxiety. We'll talk about how to realistically wean yourself off caffeine in chapter 7.

Sleep Shoe Size

Patients often ask me, "What's the right amount of sleep?" Many of us are familiar with the idea that people need between seven and nine hours of sleep. In fact, as German chronobiologist Till Roenneberg outlines in his book *Internal Time,* modern human sleep needs actually follow a bell-curve distribution, where 95 percent of the population needs between seven and nine hours of sleep.[21] Only a very small portion of the population actually functions optimally on less than seven hours of sleep,[22,23,24] even though I would estimate that about half of my fellow New Yorkers consistently get less than seven hours.

Even if you're committed to getting a good night's sleep, it's not as if we're sitting at a restaurant, staring at a menu, and selecting the number of hours of sleep we need as if it were an entrée: *Let's see here, the options are seven, eight, or nine hours of sleep. Hmmm . . . I'll have the seven, please.* We don't get to *choose* how much sleep our body needs. It's something unique to our constitution. I once heard this described as your "sleep shoe size." That is, your body has a certain number of hours of sleep it needs. If you're a size seven, then go ahead and get seven hours—you'll feel good. But if you're a size nine, then seven hours is not going to cut it. Imagine your feet were a size nine and you walked around all day in size-seven shoes . . . It would hurt! The key is to know your sleep shoe size and protect it fiercely. If you're not sure what the right amount of sleep is for your body, you may have to set aside a period of weeks where you catch up on sleep debt and allow yourself to wake up without an alarm. Track how much sleep your body takes once you're generally rested and waking up on your own. Your sleep needs will vary depending on a few factors—illness, stress, and intense exercise can all increase our need for rest. But it's a good idea to have a general sense of your body's needs, and wear the right size "shoes" every night.

A special note to my size-nine friends: I know it feels as if the size sevens and eights get more out of life because they don't need as much sleep. Make peace with the fact that you need nine hours of sleep and give yourself this time to recharge. Sleep is not a waste of time—it is pure gold. Rather than resenting your body, honor it for knowing that sleep is a worthy way of spending a chunk of life. The sooner you own your need for nine hours and get those hours consistently, the sooner much of your health— and anxiety—will settle into a state of balance and calm.

Middle Sleep

When we wake up in the middle of the night, we tend to stress about it—watching the clock and panicking about being tired the next day. But in many instances, waking up in the middle of the night is actually a normal physiological occurrence called "middle sleep." This is a break between two segments of sleep. While middle sleep is normal, exposing the brain to blue-spectrum light in the middle of the night can confuse your circadian rhythm, leading to suppressed melatonin and difficulty falling back asleep. So, the next time you wake up in the middle of the night, consider if this may be the break between two four-hour blocks of sleep. Instead of stressing, simply protect yourself from blue light—i.e., don't look at the phone—and enjoy the time without pressure to fall back asleep. Generally you can expect to get naturally sleepy after about fifteen to sixty minutes. Usually, it's our angst about being awake and our incessant clock watching that rouse us into a stress response, which squanders any hope of relaxing back into sleep. Instead, try resting with your eyes closed in the dark, trusting that this is a normal wakening in between sleep segments. Before you know it, you'll be back asleep. The fundamental paradox of sleep hygiene is that, while sleep is critical for physical and mental health, it's easier to fall asleep when we don't overthink it. We have to tell ourselves not only that getting enough sleep is important but also that we'll be OK either way.

Is It Possible to Sleep Too Much?

Here are my thoughts: No.

The body takes the amount of sleep it needs. However, needing a lot of sleep *can* be a *sign* of underlying issues. I see three common scenarios that cause "excessive sleep."

1. Most common: your body is right, society is wrong.
    i. You may simply need a lot of sleep, but our productivity-obsessed society tells us it's wrong. Are you a nine-hours-of-sleep

person (as many of my patients with anxiety are), but you second-guess it and wonder if there's something wrong with you? There's not. For you, getting less than nine hours of sleep might be contributing to your anxiety.

2. There is an underlying condition that is increasing your body's *need* for sleep (and can independently contribute to anxiety).[25] Examples include:

   i. Hypothyroidism

   ii. Depression (both the depression and the increased need for sleep may be independently caused by inflammation—more on this in chapter 8)

   iii. Chronic infections (such as Epstein-Barr virus or Lyme disease)

   iv. COVID-19 long-haul symptoms

   v. Medication side effects (e.g., atypical antipsychotics such as Abilify)

If any of these are the case, address the root cause to help recalibrate your body's need for sleep. If you're just someone who needs nine hours, get those nine hours.

TROUBLESHOOTING: SLEEP SUPPORTS

If you've tried all the solutions above, and you're still struggling to sleep, there are other easy, low-tech practices that can help, such as writing down your to-do list before bed (this effectively outsources your troubles to a piece of paper, and your mind can stop juggling and relax),[26] breathing exercises, and progressive muscle relaxation—which basically means alternately tensing and relaxing the major muscle groups of the body. Also, don't overlook the fact that anxiety at night is largely determined by how well we manage stress during the day. The more you can do to build relaxation into your daylight hours, the calmer you will feel in the dark. In that sense, every anxiety-reducing suggestion in this book helps improve your sleep.

In addition to these interventions, there are some safe sleep aids that are worth exploring:

Magnesium Glycinate: Magnesium is involved in more than six hundred biochemical reactions in your body, and supplementing with it can help with insomnia, anxiety,[27] depression,[28] migraines,[29] menstrual cramps,[30] muscle tension, and many other ailments. Most of us are deficient in magnesium because our food is grown in soil depleted of minerals. If you're eating produce just picked from mineral-rich soil in the shadow of a volcano, your magnesium levels may be good. But if you're getting your sustenance from the anemic agribusiness of the American food system, you'd likely benefit from supplementation. In addition to insomnia and anxiety, other signs that you might be deficient include headaches, fatigue, and muscle cramps.

I recommend nearly everyone supplement with 100–800 mg of magnesium glycinate at bedtime. If you develop loose stool, lower your dose. If you don't want to take magnesium in pill form, there are food sources (such as dark chocolate, pumpkin seeds, leafy greens, avocados, bananas, and almonds), or you can absorb it through your skin during a relaxing Epsom salt bath. I alternate between getting my magnesium in pill form and getting it from a soak in Epsom salts.

Weighted Blankets and Cooling Pads: I have had many patients over the years benefit from a weighted blanket, finding that it calms their nervous systems. Think of it like an embrace, or the security of a swaddle. There is preliminary evidence that it can help with both anxiety and insomnia.[31]

Cooling pads are another game changer for some. Humans sleep optimally in a cool room (around 65 degrees Fahrenheit is best[32]). This is likely because a cool room approximates the drop in temperature that occurs in the natural environment after sunset, which helps cue the sleep-hormone cascade. If you know you run hot at night, cooling pads can be programmed to make the bed feel warm and cozy as you're climbing in and then cool your bed overnight to keep your body in deep sleep.

Sleep Trackers: Sleep trackers have one significant benefit, which is that they get people to start prioritizing sleep and recognizing, firsthand, that things such as alcohol, a late bedtime, and screens in the evening *objectively* impact sleep quality, making it measurably worse. Research has shown that alcohol, in any amount, may allow you to fall asleep faster, but it also increases the chances that you will find yourself awake for the second half of the night.[33] The facts are in, but if you need a fancy device to be convinced that alcohol negatively impacts your sleep, then go for it.

Melatonin: Melatonin is not a sleep-inducing substance, but it *is* a substance that tells your body what time of day it is,[34,35] which is why taking it in pill form helps so many people sleep, since it counteracts our confusing modern light cues. Call me old-fashioned, but I'm in favor of using the actual *time of day* to tell our bodies what time of day it is. In other words, using everything in this chapter, especially strategic choices about light exposure in the morning and at night, will allow your body to produce melatonin endogenously (i.e., from within your body). I tend to believe that the pill form will never be quite as good as the exquisitely orchestrated substance our body can secrete in the perfect amount with the perfect timing in response to the appropriate cues. In other words, maybe pop some melatonin on a red-eye flight, but otherwise, let your body experience darkness in the evening and enjoy the deep restorative sleep created by your body's homegrown melatonin.

Chapter 6

Techxiety

Technology, remember, is a queer thing. It
brings you great gifts with one hand, and it
stabs you in the back with the other.

—*C. P. Snow*

On any given day, we might text, Snap, Zoom, DM, Fortnite,
and FaceTime—and though these may give us the illusion of
having diverse social interactions, none is fully nourishing our
fundamental human need for connection. Without the sensorial
context of being with people in real life—the sounds, smells,
and touches of in-person contact and the shared experience of
what's happening around us—our various screen interactions are
not scratching the actual itch of fellowship. As human beings,
we are social creatures; no matter how introverted we may be,
personal connection is a nonnegotiable need. When we're primarily drawing our sense of community from screens, however,
we can be left feeling more disconnected and anxious than supported. As recent research shows, social media use is associated

with higher rates of depression and anxiety.[1] One study found that we experience decreased mood after only twenty minutes of Facebook use.[2] There is also research demonstrating that reducing online social media use can elevate well-being.[3]

And even before the COVID-19 pandemic hit, driving everyone into their homes and onto their screens, public health experts were calling attention to what they deemed the "loneliness epidemic" in the United States; a January 2020 report from health insurer Cigna suggested that around 60 percent of American adults felt some degree of loneliness.[4] Social isolation and increased use of technology impact people of all ages, but they seem to be most detrimentally affecting Gen Zers,[5] who have practically grown up with iPhones as rattles. According to a 2019 study led by psychologist Jean Twenge, PhD, more adolescents and young adults experience depression and anxiety today than a generation ago.[6] Twenge suggests that this spike in mental suffering can be attributed, in part, to the stratospheric rise in smartphone use and social media. In 2009, for example, about half of high school seniors visited social media sites every day; today that number is closer to 85 percent.[7] And studies have elucidated how more time spent on social media puts people at risk of mental health problems.[8] Young women may be suffering the most, as Greg Lukianoff and Jonathan Haidt point out in their book *The Coddling of the American Mind*, "because they are more adversely affected by social comparisons (especially based on digitally enhanced beauty), by signals that they are being left out, and by relational aggression,"[9] all of which occur more readily on social media than "IRL" (in real life).[10]

Last, there is also evidence that in-person community—such as can be found among friends, coworkers, neighbors, and fellow members of a support group—can actually *relieve* depression and anxiety.[11] One recent study out of New Zealand demonstrated that social connectedness was a stronger and more consistent predictor of mental health than the other way around. In other words, it's not just that we isolate as a result of mental health struggles. The isolation *itself* often comes before mental illness and seems to impact our mental health negatively. And social connectedness can serve as a "social cure" for psychological ill-health.[12] In Part III, we'll explore strategies we can use to build community back into our lives.

Boundaryless Workplace

I want to remind the knowledge workers among my readers—those of us who can do our jobs from a computer or a phone—that technology has brought us the anxiety of the boundaryless workplace. Bus drivers, anesthesiologists, and baristas know in their minds and bodies when they leave the physical site of their work that they are done for the day. Many of us, though, now carry our work everywhere, literally in our pockets. The pressure on us to work all the time has more than matched the efficiencies we may have gained with technology. This has only worsened with the pandemic-induced transition to working from home, which some companies seem to be embracing as the new normal. And without the satisfaction of tangible in-person interactions with our managers and coworkers, some of us feel as if we need to justify our employment with an ever-present green dot of availability online, replacing our commute time with more work time and stretching our workday by a few hours.

While there's no returning from technology, it's up to each of us to set conscious work boundaries. Social media and smartphones are still new enough that we've barely begun to establish etiquette around how to engage with them. I once heard our situation described as if we were driving around in cars, but seatbelts had yet to be invented. For the vast majority of us, the answer to this new boundarylessness is not to move off the grid and throw away our phones but instead to pause to listen to our anxiety, reflect on what it's telling us, and set new limits around technology. In one sense, the onus is on us—if we want to have less tech-related anxiety, we need to invent our own seatbelts. By this I mean: building in unplugged breaks during the day; keeping the phone away from the dinner table and out of the bedroom at night; and trying to resist the impulse to answer every ping and message as it comes in. But as a culture, we need to collectively recognize that the workplace must also respect the energetic limits of employees. There are companies that have come up with creative solutions for surviving while treating employees humanely, such as "uninterrupted Tuesdays," "no-meeting Wednesdays," and four-day workweeks—all of which ultimately improve morale and decrease burnout and employee turnover. It would be beneficial to everyone if more companies followed suit.

SOCIAL MEDIA:
ATTEMPTING TO FILL OUR NEED TO FEEL WITNESSED

Without a personal sense of community—once found more readily in churches, in villages, and with extended family members living in our homes—we are left with an increasing need to feel witnessed through the trials and tribulations of our lives. These days, we're more likely to be siloed off in our homes; sometimes a Zoom meeting is the most social connection we have in a day. Conveniently, over the last couple of decades, social media has stepped in to become our surrogate village. Now when we have a baby,

take a vacation, or buy a latte, we post a photo as a rite of passage—
and to feel witnessed. Millennials even have a catchphrase for this:
"Pics or it didn't happen." This is how we allow others to share
in our experiences and milestones, reinforcing not only that they
actually occurred but also that we matter. Yet feeling observed by
this intangible surrogate village is not wholly satisfactory. It's akin
to the way artificial sugar hits our brain—we think we're experiencing something sweet, until the aftertaste makes us realize that
was just a chemical tricking our brain. Our digital communities
offer a mere shell of being sustained by others in person, and there
is something deep within us that knows it's inadequate.

As a social species we are wired to be in personal relation with
others. This relates to our origins—we were never the fastest or the
strongest animals on the savanna. It's hypothesized that it was our
ability to cooperate with one another that allowed us to triumph
as a species. And this comes with a hardwired need for connection. In fact, community is such a biological imperative that social
disconnection feels like physical pain to the brain. As Matthew D.
Lieberman, a social psychologist at the University of California,
Los Angeles, proposes in his book *Social*, this is an evolutionary
adaptation to promote survival and reproduction. "The pain of
social loss and the ways that an audience's laughter can influence
us are no accidents," Lieberman writes. "To the extent that we
can characterize evolution as designing our modern brains, this
is what our brains were wired for: reaching out to and interacting
with others. These are design features, not flaws. These social ad-
aptations are central to making us the most successful species on
earth."[13] While social media appears to correct for the isolation
of modern life, it effectively builds more walls between us. Our
time spent connecting with the phone or the computer comes
at a cost—the opportunity cost of time we're not seeking out

in-person interaction. Our online lives ultimately hinder our ability to truly connect and leave us unsatisfied and anxious.

Positioned for Anxiety

Staring at screens may be telling our brains we're anxious. The neck and shoulder position we hold during extended hours at the computer and staring at our phones impacts blood flow to the brain[14] and puts tension in the critical musculature of the neck, upper back, and jaw, all of which are connected with our sympathetic nervous system. The fixed position of our eyes on the screen and the clenching of the jaw and trapezius muscles signal to our brains that we're in a stressful situation—whether we are or not. So, a perfectly amiable video conversation may be more physiologically anxiety provoking than we realize. When we focus on the screen, at times our eyes widen similarly to the way they would in a fear state. Be aware of the alignment of your neck and the softness of your gaze when you work at your computer or look at your phone. Can you feel muscle tension develop as you tilt your neck forward? Are your eyes straining? If so, it's worth creating better ergonomics when working, taking periodic breaks to rest your eyes and, on occasion, putting the technology aside to take a few unplugged minutes outside.

USE TECHNOLOGY; DON'T LET IT USE YOU

It's been said that we live in the attention economy. That is to say, our attention is *the commodity* being competed for by media and advertisers. And these companies—whose profits rely on commandeering our eyeballs—have thoroughly done their homework. "The thought process that went into building these applications . . . was all about: 'How do we consume as much of your time and conscious attention as possible?'" Sean Parker, the founding president of Facebook, confessed in 2017 at an Axios event.[15] These news and social media companies are all keenly aware of the neuroscience and behavioral psychology that goes into their success; they know how to exploit our fear response as well as entice the reward circuitry in the brain that lights up at a quick hit of validation. They know that getting a "like" on Instagram triggers the release of dopamine, a neurotransmitter involved in reward, at unpredictable intervals—much like the way we feel when we get a match on a slot machine—leaving us wanting more.[16] Meanwhile, media companies are also cognizant of the fact that if they offer headlines that instill fear and anxiety, uncertainty, or doubt, or leave us feeling inadequate, we'll tune in. Not to mention that controversy makes us rubberneck. The bottom line is that while their revenue increases, *our* mental health is the collateral damage. My best advice to counter this grim truth actually comes from an unlikely source—Sean Parker himself. "I use these platforms," as he once put it, "I just don't let them use me."[17] We need to make very intentional choices about when and how we give our attention away to news and social media in order to protect our psychological well-being.

Many of my patients are profoundly affected by the media they consume. They are glued to the twenty-four-hour news cycle or destructively comparing their real lives to the curated highlight reels they see on Instagram—freely offering up an increasingly large share of their attention and, in return, suffering from debilitating doses of anxiety. One such patient, Aisha, a thirty-six-year-old woman who worked as a magazine editor, devoted entire sessions with me to talking about how on edge she would feel when she interacted with Twitter or Instagram. "I'm trying to stay current, and I care deeply about these issues, but I'm starting to wonder if I'm addicted to the news. Sometimes I can't think about anything else, and I check my phone constantly." She added that when she posted on social media about something she cared about, the comment section left her feeling "misunderstood" and "attacked." Aisha's difficulties in this regard were heightened by the fact that, to be successful in her career, it was advantageous to keep up an active social media presence.

I advised her to be brutally honest with herself about how much she was expected to be on social media for work. My hunch was that she was telling herself she needed to do it far more than was truly required. "What characterizes an addiction? Quite simply this: you no longer feel that you have the choice to stop," as the spiritual teacher and author Eckhart Tolle succinctly puts it in his book *The Power of Now*. "It also gives you a false sense of pleasure, pleasure that invariably turns to pain."[18] I advised Aisha to pause and reflect on the potential consequences of her actions *before* she opened Twitter and got sucked in.

I also encouraged Aisha to make *conscious* choices when navigating the information landscape—choosing wisely who gets to tell her the news and how often. To be clear, responsible journalism is *not* the enemy of the people. Now more than ever, we need our professional investigative truth tellers out there finding and publishing the truth. Unfortunately, in this so-called age of information, the story that's a click away often leads to imbalance and fear, instigating the stress response and stoking the fires of anxiety.

Aisha experimented with deleting apps she didn't need for work and limiting her news to a few trusted sources and a few check-ins per day. She also agreed to a technology curfew—powering down all electronics an hour before bed and, crucially, keeping her phone out of the bedroom overnight. She now uses her extra time in the evening to take a bath or read a book instead. Aisha's anxiety improved, and it turned out that she was still effective in her work—arguably more effective—even though she wasn't plugged into the news at all hours. These shifts allowed us to investigate, with a little more distance and perspective, her feelings about the contentious tone so pervasive in much of social media.

CANCEL CULTURE

Another way that technology stirs up our anxiety is the climate of cancellation that occurs on our social media forums. While cancel culture is accountability culture—offering a necessary reckoning for bad behavior that helps us grow and improve as individuals and as a society—it can also veer into more detrimental territory. As we scroll through social media on our phones for hours every day, cancel culture means we are bathing in criticism and relational aggression, which—given that we are wired to feel safe when supported by community can leave us feeling anxious and quietly questioning our basic sense of security. While there are important and long-overdue conversations happening online, we need to balance our awareness and participation with our own mental health—after all, we can best fight to make the world better when we are feeling strong and well. As activist and co-founder of Campaign Zero Brittany N. Packnett Cunningham puts it, "We need rested warriors."[19] When you notice technology is leaving you strung out, sometimes the right choice for your mental well-being is to step back and take a breath.

In general, I advise all my patients to navigate the information buffet in a way that leaves them nourished rather than sick. Just like the way the food we put into our bodies can impact how we feel, what we watch, read, and hear can alter the state of the nervous system and create a ginned-up sense of doubt and insecurity—otherwise known as false anxiety. In other words, technology is a source of avoidable anxiety hiding in plain sight.

Chapter 7

Food for Thought

We have come to this strange cultural
moment where food is both tool and weapon.

—*Michael W. Twitty,* The Cooking Gene: A Journey
Through African American Culinary History in the
Old South

We are living in the wake of decades of misleading and contradictory nutrition recommendations—first being told to eat
low fat, then advised to turn our attention to eating low carb;
the virtuous brunch order has gone from an egg-white omelet
to a grass-fed steak; butter, once an instant heart attack, is now
a healthy addition to our morning coffee. Figuring out what to
eat would make any reasonable person's head spin and can be a
source of anxiety in and of itself. Meanwhile, rather than pass-
ing on traditional food wisdom from generation to generation,
as was the case with indigenous tribes, American food "wisdom"
is a pyramid of serving-size recommendations heavily influenced
by industry funding.[1] So we are left to figure out for ourselves

how to eat. In an attempt to both take control and take good care of their bodies, many of my patients veer toward "clean eating." I'm not opposed to green juice or smoothies, but it's worth noting that you could spend your days dining on chia-seed pudding and Instagram-worthy oat-milk matcha lattes, and while you'd be eating perfectly "clean," you might also find yourself feeling malnourished and anxious. You can, in fact, eat completely *clean* and still be missing the nutrition needed to feel *well*.

When it comes to eating for our mental health, we need to get back to balance. And while there may be merit to eating clean or paleo or low carb, these types of labels can be destructive. When we get too attached to a particular style of eating, we walk a delicate tightrope between paying attention to what we eat and falling prey to obsessive thinking about food, which is not only conducive to anxiety but can also lay the groundwork for eating disorders.

I have some personal experience with this issue. In my first year of medical school, I started binge eating. A few weeks before the bingeing began, I had started restricting my food, in a misguided attempt to control my body in a time when everything else in my life felt unmoored. Looking back, I believe a few factors contributed to my slide into bingeing: food restriction itself pushed my brain to obsess about food, which is a biologically adaptive response to a caloric deficit, designed to urge a person to seek out food to survive. I also felt disconnected and out of alignment with my purpose throughout much of med school, and I used emotional eating as an unconscious attempt to self-soothe, yearning to fill up my sense of emptiness and isolation with food. But I suspect that the most important (yet least understood) driver of my binge eating was addiction—and food was my drug. This addiction expressed itself as a certain feeling in my gut—a bloated, out-of-control craving deep in my intestines—that, when combined with the stress and loneliness I felt, drove me to binge. I would devour pizza, cookies, grilled cheese sandwiches, and ice cream, approaching food the way an addict in withdrawal might return to their drug of choice and overdose.

I feel sorrow when I think about this time in my life. Bingeing was a secretive, time-consuming, shame-filled, physically uncomfortable, and surprisingly expensive behavior (eating large quantities of food requires a lot of trips to the grocery store). I began to gain weight with reckless abandon. My knees hurt from the added pounds they had to carry around. I eventually found a therapist with a specialty in treating eating disorders, and slowly I recovered. I count myself lucky for recognizing that I needed help and for being able to access the right support, particularly as treatment is still so stigmatized, inaccessible, and difficult to navigate, leaving many to suffer in silence for years.

FOOD ADDICTION

No addiction is "easier" to manage than any other, but unlike some addictive substances, food is uniquely challenging in that quitting it entirely is not an option. We must interact with our drug multiple times a day, which makes it hard to establish recovery. As bestselling author, lecturer, and research professor at the University of Houston Brené Brown put it: "I once heard someone say, 'Abstinence-based recovery is like living with a caged, raging tiger in your living room. If you open the door for any reason, you know it will kill you. The non-abstinence-based addictions are the same, but you have to open the door to that cage three times a day.'"[2] A sizable number of my patients with anxiety struggle with binge eating. In trying to help them, it occurred to me that, while my patients can't abstain from food, they *can* abstain from *drug-like foods.*

Yet a cornerstone of eating disorder orthodoxy is that *no* food is off-limits, as long as you take a measured approach. "Everything in moderation," the books, therapists, and internet tell us. But abstaining from gluten, dairy, sugar, and processed foods has allowed many of my patients to circumvent their triggers enough to stop bingeing. Although to a typical therapist, this restriction might look like just another variety of eating disorder, this has been the path to recovery for many of my patients. There is in fact a scientific explanation for why this happens: gluten breaks down into something called gluteomorphin, and dairy breaks down into casomorphin. Do you hear that root—*morphin*? Mac and cheese not only is delicious, but it also behaves like a baby dose of a morphine-like chemical that makes us want more.[3,4] Sugar is excitatory to the brain, and processed foods are engineered to exploit neural reward circuitry in the brain. Betcha Can't Eat Just One, as the Lay's potato chips slogan boasts. But the joke is on us—*we literally can't,* because it's engineered to be addictive.

When my patients are able to break this addiction, they then, over time, come to relearn the feeling of satiety as well as develop a sense of freedom with food. Better yet, many of them are eventually able to return to their trigger foods without precipitating a binge. Furthermore, when my patients stop bingeing, their moods stabilize, digestion normalizes, and they're able to resume going out to dinner with friends. True recovery always involves healing on a deeper, psychospiritual level as well—reclaiming self-acceptance and positive social interaction and reconnecting to meaning. But the abstinence technique can offer a crucial exit ramp, allowing people to climb out of food addiction long enough to engage in the psychospiritual work needed to fully understand and heal the behavior. So, I would like to make the radical suggestion that some eating disorders—binge eating, bulimia, and even certain forms of anorexia—have addiction to drug-like foods at their root, and abstinence is the path out.

The Clean Eating Disorder

After years of helping patients discover which foods were behaving like drugs in their bodies, I began to believe that the key to solving any eating disorder, as well as helping with depression and anxiety, was to wean people off eating drug-like foods and shift them to eating real foods. I used this approach in my practice—and continue to use it—with great success.

But I also discovered a pitfall along the way: there are times when this approach can actually *cause* a different type of disordered eating called orthorexia, which is an obsession with eating the "right" way. This condition has sadly become somewhat common in the wellness community. In paying very close attention to the parameters for eating well, some of us begin to hyperfocus on the restrictions. When this occurs, people start to become fixated on eating perfectly all the time—they become preoccupied with meal prep and start declining dinner-party invitations—and they become anxious when they can't control the circumstances of a meal. Their lives grow smaller and more rigid, they lose social connections, and though they may hit some of their health goals—less bloating, less bingeing, improved insulin sensitivity, a healthier gut—their anxiety levels *increase*.

British actress and activist Jameela Jamil expressed the problems inherent in wellness culture's obsession with food. In a social media post, she wrote: "Diet culture . . . was my slippery slope to losing all sense of reality. . . . [L]ook out for signs of associating food with guilt, shame, anger or failure. Listen to the words you say to and about your body in the mirror. . . . Would you ever tell someone you love and respect that they need to meet the same body goals to be allowed to feel good and confident? Is it ok if your body doesn't end up obeying your fantasy because maybe that's just not how you're supposed to be built? . . . Why is what our clever body wants and needs not a priority?"[5]

I have now learned to always follow any discussion of food choices with my patients with a reminder to keep their relationship with food free and easy, based on trust rather than fear. It's important not to deny ourselves the multifaceted pleasures of food, which are, themselves, powerful medicine for anxiety. For example, if in the process of eating in a way that best serves your body, you've begun avoiding social engagements, then perhaps you've gone too far. I would argue that a life-affirming social connection over a meal is better for your health than any amount of healthy eating, so avoiding dinners with friends to "eat clean" is counter-therapeutic. I also offer my patients troubleshooting tips—such as eating before you go out or bringing along a dish to a dinner party—for navigating the inevitable challenges that come up when they start to eat in a new way that doesn't easily jibe with the world around them.

Additionally, when we think about the dangers of fearing food, it's important to remember that our brain is forever *learning* (because that's what brains do), and it learns anxiety in much the same way that it learns algebra. To this end, be aware that anything that suggests you need protecting increases anxiety. When we navigate the nutrition landscape from a place of *fearing food*, even if this might result in our eating better foods, it has a net effect of increasing anxiety. We need to learn the art of eating well in a way that doesn't feel like deprivation or a threat. If you can create a sense of safety and abundance around eating—taking pleasure in the real foods that help our bodies also fend off anxiety—you'll be able to navigate nutrition *and* life with more ease.

A NUANCED TAKE ON BODY POSITIVITY

My patient Valerie, who came to see me for help with her depression and anxiety, is a champion of the body positivity movement. In our initial visit, the moment I hinted that I might suggest diet changes, she went on a persuasive tirade against diet culture and the "insufferable influencers on Instagram" who promote it. Only moments later, however, she went on to tell me about the issues she was having with her "lady parts," adding that she experiences irregular and extremely heavy periods, with painful cramps and headaches that make it hard for her to work and go about her life for a week of every month.

So, one minute she was saying, *Shut up, diet fanatics, stop shaming me for eating junk food*, and the next she was telling me about how her body is breaking down, causing great suffering. What dawned on me is that Valerie didn't see a connection between what she eats and how her body functions.

My hope is that, culturally, we're coming to better appreciate this connection: our food choices contribute to states of imbalance in our bodies that can create physical and mental suffering. To put a fine point on it: what we eat impacts how we *feel*, and that's why this matters. This encounter with Valerie prompted me to rethink my views on the body positivity movement. This movement was encouragingly telling Valerie, *Eat whatever you want, and don't suffer to please the patriarchy*—but this approach was ultimately creating *more suffering* than ease in her life; it was standing in the way of her goals of feeling less depressed and not losing a week of every month to pain and heavy bleeding. Body positivity, like so many things in life, should be viewed with a nuanced, both/and understanding.

For the most part, I resoundingly agree with the principles of the body positivity movement. As Sonya Renee Taylor puts it in her revolutionary book *The Body Is Not an Apology: The Power of Radical Self-Love*, "There is no wrong kind of body." I am in favor of celebrating all body types, rejecting diet culture, discarding the presiding racist and sexist ideals about what a body should look like (and the idea that what a body looks like *should even matter at all*), recognizing that we can't judge someone's health by a dress size or a number on the scale, and pushing back against the fat shaming that occurs at job interviews and in the doctor's office and . . . well, *everywhere*. Body positivity is an important movement speaking truth to patriarchal ideals that have conditioned many of us to turn on our bodies.

However, I also have firsthand experience with being addicted to processed foods and being so physically out of balance that, like Valerie, it was difficult for me to live a fulfilling life. We need to reconcile the principles of the body positivity mindset with the awareness that what we eat and the state of our physical health impacts our mood and how we function in our lives.

While there is indeed a patriarchy selling us diet culture and capitalizing on body shame, there is also an equally powerful patriarchy lining its pockets by selling us addictive foods that derange our appetite and metabolism. And the body positivity movement has largely overlooked the Big Food patriarchy. So, here's what I tell my patients: Make sure you bring into focus *all* the patriarchies that are standing between you and a fulfilling life. It's not exclusively the one telling us to be thin.

After all, what we eat impacts how long we live,[6,7] how *well* we live,[8] and whether our body can be a vibrant foundation for the work we're here to do on earth. Obsessing about food and the number on the scale holds us back from carrying out our goals and enjoying our lives, but so, too, does poor mental and physical health.

I don't blame individuals for straying from healthy eating. I do, however, blame the industry that knowingly sells us addictive food. I blame the manufacturers that release endocrine-disrupting chemicals into our water supply and convince us to apply these chemicals directly to our skin. These exposures create hormonal imbalance and disturb our metabolism. I blame the scientists who agreed to be paid off to distort the scientific consensus on sugar and saturated fat.[9] I blame the government that has enabled this total lack of regulation and sold out our health to corporate interests at every juncture. So, yes, let's rebel against diet culture, but not against our own needs. To feel well, we need to eat well, but for none of the reasons that diet culture tells us to. How we feed ourselves should be an act not of self-negation but of radical self-love. And navigating the food landscape from a place of self-love means discerning when easy food causes a more difficult life.

When Valerie ultimately figured out that her body doesn't tolerate dairy and that sugar was affecting her mood and hormones, she reduced both considerably. Her depression and anxiety improved, as did her physical health, including her previously undiagnosed polycystic ovary syndrome and endometriosis. The lesson: let your food choices be an intimate decision-making process that happens within the boardroom of your own heart. The patriarchies of diet culture *and* Big Food have no place at that table.

HANXIETY

In my practice, I start with the assumption that anxiety is a blood sugar issue until proven otherwise. I am not being dismissive of people's very real suffering, nor am I implying that everyone with anxiety is diabetic. The truth is that blood sugar is not binary—you're not either diabetic or perfectly healthy. For many of us, our bodies are operating somewhere along a spectrum of dysglycemia, in which a subtle, subclinical impairment with blood sugar regulation causes us to swing up and down throughout the day, with every blood sugar crash generating a stress response.[10] Given that the modern diet is so blood sugar–destabilizing, these stress responses are at the root of much of the anxiety I see in my practice. And I've found adjustments to blood sugar to be among the most immediate and effective salves for anxiety. If you're familiar with the experience of being "hangry," then you probably also get *hanxious*, meaning you get anxious when your blood sugar is low.[11] If this rings a bell, it's worth taking a closer look at your diet and the role blood sugar may play in your mood. Even if your blood tests are normal and your doctor has never mentioned diabetes to you, I'll wager that stabilizing your blood sugar will help your anxiety.

Throughout human evolution, getting enough to eat has been a life-and-death issue, so the human body has a series of checks and balances to ensure blood sugar security. Our bodies store sugar in the form of starch, called glycogen. When our blood sugar is low, a cascade of events is triggered. The adrenal glands release adrenaline and cortisol—the stress hormones—telling our liver to break down the glycogen into glucose and release it into the bloodstream. The adrenaline and cortisol also create a sense of urgency to seek out more food—which, in modern times, translates to the 3:00 p.m. hunt for snacks. This bodily system gets the job done—it puts glucose back into the bloodstream when blood sugar is low and motivates us to forage for more food. The only catch is that it also creates a five-alarm fire in the body. The body triggers a stress response to accomplish these goals, and that stress response can feel identical to anxiety. These blood sugar crashes are a largely avoidable cause of anxiety. So, if you suffer from hanxiety, you might find that—difficult as it can be to avoid sugar—certain dietary changes are worth it for you.

Take my twenty-eight-year-old patient Priya, for example. She had frequent panic attacks for years. When we began working together and started to examine the pattern of her panic, we noticed that she would fairly reliably experience panic attacks after eating something particularly sweet or after she missed a meal altogether. Over several weeks of observation, it became apparent that low blood sugar was triggering panic for Priya. Clearly, we needed to implement a treatment plan that included keeping her blood sugar stable. I reassured her that there was a definitive solution to this problem—and an effective hack. I recommended, first, that she eat regularly scheduled meals of blood sugar–stabilizing real food, focusing on well-sourced protein, starchy tubers like sweet potatoes, and vegetables, all prepared with healthy fats, while minimizing sugar and refined carbohydrates. As this was a 180-degree departure from Priya's typical diet of skipping breakfast and having a Frappuccino and candy in place of lunch and nachos for dinner, a look of trepidation came across her face as I spoke. So, I told her we could start with the same simple hack I use for people who experience blood sugar crashes while sleeping: eat a spoonful of almond butter or coconut oil at regular intervals throughout the day to keep her blood sugar stable. Despite looking at me as if I had two heads, Priya agreed to give it a try. At our next session, she reported that she was eating a spoonful of almond butter at around 11:00 a.m. and 3:00 p.m. and at bedtime. The kicker: her panic attacks are much less frequent and tend to occur only when she has fallen out of the almond butter routine.

Both Priya and I were thrilled with this immediate triumph, but I've had plenty of other patients who have struggled for longer and needed to implement more substantial changes to their diets. (To be fair, I also counseled Priya to continue altering her diet, even after her anxiety abated, just for the sake of her general health and well-being.) Sometimes the path out of false anxiety requires a more concerted look at added sugar in the diet. We all know that when we take that first bite of something sweet, it can instantly unleash the "Sugar Dragon . . . that fire-breathing beast clinging to your back, roaring loudly in your ear, demanding its next fix," as Melissa Urban, cofounder and CEO of Whole30, describes it. She recognizes that it's hard to come halfway off sugar, noting that "the best way to slay the Sugar Dragon is to starve it."[12] I have also found this to be true for my patients. Once they unleash the dragon, they find themselves continuing to crave sugar. Once they're fully off sugar for a week or two, however, the cravings subside.

But I get it: quitting sugar is no easy feat. Some of my patients find that the spoonfuls of almond butter or coconut oil help get them through that difficult first week. Many find that they simply have to sweat it out and get through the withdrawal period, and then after about a week off sugar, they're free— the cravings are gone, and they feel much better, physically *and* emotionally. The dragon has been slain.

If you only make one dietary adjustment in the name of curbing your anxiety, I would suggest addressing your relationship with sugar. Do whatever you can to keep your blood sugar stable, and consider reducing your sugar intake overall.

YOUR BRAIN ON CAFFEINE

There are two things nearly all my patients have in common: anxiety and coffee. If you share this duo, it's time to take an honest look at the role caffeine plays in anxiety—even beyond its impact on sleep. "I once had a patient come to me with severe recurrent anxiety," writes pseudonymous psychiatrist Scott Alexander on his blog *Slate Star Codex*. "I asked her how much coffee she drank, and she said about twenty cups per day. Suffice it to say this was not a Dr. House–caliber medical mystery."[13] It's a tongue-in-cheek illustration of how obvious this connection should be—and yet how often we overlook it.

To be clear, caffeine is not inherently bad. It's safe, enjoyable, and even potentially beneficial. Coffee is a source of magnesium and antioxidants, and it's associated with a decreased risk of Parkinson's disease[14] and type 2 diabetes.[15] Tea is also packed with antioxidants and beneficial polyphenols. However, just as with sugary food, caffeine can trigger the release of cortisol,[16,17] which can feel identical to anxiety. If you happen to suffer from generalized anxiety, panic attacks, or social anxiety and you consume coffee, tea, soda, or energy drinks, caffeine very likely plays a role in your symptoms.[18]

For susceptible individuals, caffeine can behave as an anxiogenic drug, meaning it can precipitate a stress response and induce anxiety. I see this occur in particular in people who are slow metabolizers of caffeine, which can be determined by a genetic test or simply by noticing that caffeine's effects take a long time to resolve in your body. When we ingest caffeine, it promotes the release of cortisol, which revs up the sympathetic (fight-or-flight) response. In other words, caffeine makes our nervous system ready for a fight. Then, if we introduce a stressor—say, a difficult commute or an unsettling work email—caffeine causes us to have a more pronounced reaction to that stress. Before you know it, your heart is pounding, your hands are tremulous, and your whole body feels shaky and amped. Or perhaps your mind gets tripped too readily into a ruminative spiral. Additionally, if you take an antianxiety medication *and* drink coffee, as I find many of my patients do, be aware that you're taking one drug that *gives* you anxiety and another that *treats* that anxiety. What if you just stopped taking the drug causing the anxiety in the first place?

Yeah, right, you say, *just the thought of quitting coffee is giving me anxiety.* But stay with me. I know drinking coffee has become a favorite cultural ritual; sometimes it feels like the only reliable joy in our day, our one true friend in the world. But remember that caffeine feels so good, in part, because it's the antidote to its own withdrawal—we wake up with caffeine withdrawal, and then coffee gets the credit for being the salvation to the very problem it created! Fortunately, there are ways to wean ourselves off caffeine without too much suffering or sacrifice.

That said, if I've convinced you to experiment with caffeine sobriety, please do *not* decide to go cold turkey tomorrow. I strongly suggest you make any changes to caffeine consumption *gradually*. Caffeine is a real drug with a real withdrawal. To avoid headaches, irritability, and fatigue, taper off slowly over the course of several weeks. Go from a few cups of coffee daily to one, then to half-caf, then to black tea, then to green tea. Eventually, you'll be down to a few sips of green tea, and from there you can transition to caffeine-free herbal tea. You may still have a few fuzzy days where you're getting your bearings, but with time, things will stabilize. You will be your same energetic, productive self, but without the highs and lows of caffeine, not to mention the daily outlay of $4.95 to your local coffee shop. Most importantly, decreasing or eliminating caffeine can significantly reduce anxiety. As neuroscientist Judson Brewer writes in *Unwinding Anxiety*, "The only sustainable way to change a habit is to update its reward value."[19] Notice the ways caffeine contributes to your anxiety, and observe how you feel less anxious once you're finally off; this can update the "reward value" of coffee and help reinforce the habit change. Still, for those who like the idea of living with less anxiety but truly can't imagine life without the smell, taste, and ritual of coffee, I have one word: *decaf.*

BOOZE AND ANXIETY: A LOVE AFFAIR

Humans have long used alcohol to self-medicate for anxiety, and for good reason: it works. In the short run, at least. As with benzodiazepine medications, such as Xanax, alcohol modulates the activity of the neurotransmitter GABA in our brains. When we drink, alcohol acts on GABA receptors, feeling like a rush of GABA in our synapses, and this feels good, pleasant, relaxing . . . Suddenly the things we were so wound up about no longer feel like such a big deal, and for a brief, sweet moment, we feel easygoing and confident, and it seems as if everything is going to be OK. If only that were the end of the story, it truly would be a love affair for the ages.

But, as we now know, the body doesn't really care whether or not we're relaxed; it just wants us to survive. So, when we've had a glass or two of wine, the body becomes aware that if, say, a leopard came around the corner, we'd be too buzzed to care. The brain then goes to great lengths to restore homeostasis—which it does by reabsorbing GABA and converting it into glutamate, an excitatory neurotransmitter.[20] After this, you may have GABA flowing in your synapses, but your brain can't *hear* it. And the resulting feeling? Anxiety. Alcohol chills us out temporarily, but in the end it leaves us feeling more anxious than we were to begin with. And this effect can accumulate over the long term, so it's easy to see how alcohol creates the need for itself, looping us into a vicious cycle.

Another issue with alcohol is the way it plays into our desire to numb and escape the difficult moments in our lives, which can have a detrimental effect on our ability to be awake for the full range of human experiences, to stay attuned to our inner truth, and to properly process trauma, stress, and grief.

While we have all been indoctrinated with the idea that alcohol is "heart healthy" and even an empowering choice, recent research has uncovered that alcohol is bad for our health in any amount, putting us at increased risk of cancer and dementia.[21] And when it comes to mental health, alcohol's impact on GABA exacerbates anxiety in any amount.[22,23]

If you lean on alcohol as a way of mollifying your anxiety, there is no blame or shame here. But if nobody else is going to tell you this, I feel a responsibility to let you know: it's probably making the situation worse in the long run. It would be better to give your brain a chance to build its GABA activity back up so you can feel at ease even when you don't have a buzz. This can be achieved through nutrition, yoga, meditation, breathing exercises, and gut healing (specifically repleting the beneficial bacteria that help manufacture GABA) and by avoiding substances that adversely impact GABA activity, such as alcohol. (Benzodiazepines also negatively impact GABA signaling in the long term.) Generally, I am proposing that each of us bring conscious reflection to the role alcohol plays in our overall well-being and to whether making different choices will offer some relief from anxiety.

OVERFED AND UNDERNOURISHED

In the United States, our relationship with food has become so fraught that we have nearly forgotten the connection between food and nourishment. The fact is that the functioning of our brain rests on whether our food provides the necessary raw materials. When we're well nourished, we *feel good*.

Certain foods and nutrients have direct impacts on our neurochemistry and our anxiety. For one, neurotransmitters such as serotonin and GABA, which help us feel stable and calm, are not manufactured from air—our bodies build them using nutrients from food, such as the tryptophan in turkey and the glycine in bone broth. In addition, our bodies are constantly surveying our nutritional status and deciding whether we have "enough." My belief is that when our bodies detect that we're missing vital nutrients, this sensation triggers a feeling of scarcity, urgency, and unease, motivating us to forage and leaving us feeling anxious until we have enough. And then, of course, there's the fact that sugar and inflammatory foods directly cause a stress response, which can feel synonymous with anxiety.[24,25]

To get all the necessary nutrients from our diet, we need to eat a wide variety of nutrient-dense foods. So, let's take a closer look at how we can eat to support our well-being.

WHAT TO EAT FOR BETTER MENTAL HEALTH

Here are the components of your new plate: in general, about a quarter of your plate should be a well-sourced protein, a quarter should be starch, and half should be vegetables, with healthy fats throughout.

The one thing almost all nutrition experts seem to agree on is that vegetables are key to good health. By offering abundant vitamins, minerals, and antioxidants, vegetables support brain function and help with anxiety. So, eat all the vegetables, and plenty of them. Let vegetables be at least half your plate and the centerpiece of each and every meal. In the warmer months, aim to eat more raw vegetables; in the colder months, go for cooked and stewed vegetables. Prepare them with healthy fat, such as olive oil, avocado oil, or ghee (clarified butter). If it's in your budget, try to buy organic whenever possible, especially for skinless veggies. (And use the "dirty dozen" and "clean fifteen" from the Environmental Working Group to navigate these decisions with an eye toward cutting costs. If you're not familiar with this, these are the twelve fruits and vegetables that carry the most pesticide burden and the fifteen that you can buy conventional without too much cost to your health[26].)

When adding protein—which provides the necessary building blocks for peptide neurotransmitters, such as serotonin—to your diet, aim to eat a variety of well-sourced, pasture-raised meat and poultry as well as wild, small, cold-water fatty fish, such as sardines, anchovies, arctic char, and salmon. To ensure you get a wide range of nutrients, eat a variety of animals. If you see them on the menu, opt for game meats—the less mainstream the meat, the less likely it's a product of large agribusiness.

We currently have a cultural obsession with *lean protein*, so skinless chicken breast seems to be everybody's go-to. Chicken can be part of a well-rounded diet, but it should not be your only protein source, because it doesn't meet all your nutritional needs (and when you do eat chicken, don't compromise on the quality—go for pasture-raised chicken, free of antibiotics). Aim to eat poultry about once a week and red meat and fish the rest of the time. Even better, become well acquainted with your body's cravings, and learn to reach for the protein your body is telling you it needs.

In addition to eating a variety of animals, I also encourage you to eat *every part* of the animal—nose to tail. Organ meats have fallen out of favor in the West because we're so focused on eating muscle meats. But organ meats are uniquely nutrient-dense, so it's worth making a point of seeking them out. Admittedly, it can be tricky to figure out how to incorporate organ meat into a modern American diet. The way I make it work is by purchasing pastured chicken liver pâté from my butcher. Chicken liver is a good source of zinc, copper, manganese, vitamins A and C, B vitamins, iron, phosphorus, and selenium.[27] In other words, it's Mother Nature's multivitamin. If you can throw back a spoonful of pâté every few days, it'll go a long way toward meeting your nutritional needs.

When it comes to gleaning the nutritional benefits of meat for brain health, it doesn't require eating a steak the size of your head. Think of meat more as a condiment than as a centerpiece. And if eating meat does not jibe with your ethical principles, my vegetarian standbys are any combination of rice and beans (which form a complete protein when eaten together), eggs, and full-fat dairy (if tolerated).

Reconsider Meat

You may find this surprising, but I often recommend my patients incorporate *more* red meat into their diets. As a former vegetarian, I don't approach this advice lightly. I appreciate and respect the deeply held choices people make when they feed themselves. In fact, I reconsider the necessity of meat in my own diet on a daily basis. But I can't deny what I've witnessed with my patients and in my own body—that incorporating meat, in particular red meat, is sometimes a critical step toward healing mood and anxiety. Meat is a nutrient-dense food and often the most bioavailable way to get certain nutrients, such as iron and zinc, in adequate quantities.[28] I also believe that meat can benefit our health in less tangible or readily measurable ways. In Chinese medicine, for instance, meats, stews, and bone broths are understood to "build blood" and support critical needs such as kidney qi, which, when deficient, can lead to hair loss, weakened stamina, cold intolerance, knee pain, and a tendency toward fear.

With respect to optimizing physical health, I usually recommend adopting a diet that approximates that of your ancestors. What that diet looks like will vary depending on where you descended from geographically. It can range from a semivegetarian diet with small amounts of fish to a heartier diet of red meat and tubers. For many of us, the inclusion of some amount of animal food—even in very small amounts—can go a long way toward rounding out our nutrition and addressing the root cause of anxiety.

I know these recommendations will raise plenty of questions, and even stir up ire, among those who have made a choice to eat a vegetarian or vegan diet. So, let me clarify: if ethics and animal rights are your priority, eating meat may be out of the question for you. I honor that fully. This suggestion is really aimed at anybody eating a vegan or vegetarian diet because you think it's *healthier*. If that's the case, I would urge you to reconsider. In fact, after years of telling us that red meat is unhealthy, academic nutrition scientists quietly admitted in 2019 that research does not support that claim.[29] So, if it feels comfortable for you, try adding in beef bone broth, chicken soup, or even some chicken liver pâté, and see if your health and anxiety improve. And for everyone consuming meat, remain conscious about animal welfare, bring reverence to the act of eating meat, and avoid factory farmed meat entirely, for ethical, environmental, *and* health reasons.

CARBS: FRIEND OR FOE?

As with most nutrition advice, low-carb diets are constantly cycling in and out of vogue. But, as always, there is no one-size-fits-all nutritional recommendation on this issue. More often than not, though, I have found that my patients who struggle with anxiety have found relief by *allowing* themselves to consume carbs rather than eschewing them. This effect has been most pronounced among women of reproductive age. Just as with fat (which we'll talk about in a minute), we need to have a more nuanced cultural conversation about which carbs are healthy and which are not.

It's difficult to communicate this distinction without creating confusion. I often encourage patients to minimize their consumption of *refined carbohydrates*, as these are generally inflammatory[30] and blood sugar destabilizing, but somehow this advice can get misconstrued to mean: avoid carbohydrates altogether. Most of us do well with carbs, and I recommend that you fill the remaining quarter of your plate with them. In fact, I think a lot of people, especially health-conscious folks, actually need to *increase* their carbohydrate intake to signal to their bodies that they're not, in fact, living in a time of famine, and thereby quiet their bodies' stress response. To be clear, this does not mean leaning into pasta, bread, crackers, and baked goods. Those are *refined* carbohydrates, which are inflammatory, put us on a blood sugar roller coaster,[31] and can be damaging to our health, contributing to diabetes,[32] obesity,[33] dementia,[34] heart disease,[35,36] digestive issues,[37] a shorter life span,[38] and anxiety.[39]

But carbs from starchy vegetables—such as sweet potatoes, white potatoes, plantains, squashes, taro, and yuca—are not only OK but often beneficial and necessary for staving off anxiety. It takes more time for our bodies to digest and absorb the carbohydrates from starchy tubers compared with refined carbs, offering us a steady supply of blood sugar without the spike and crash. And these carbohydrate sources contain something called resistant starch, which can serve as the food for beneficial bacteria in our guts—which we will explore further in the next chapter—and thus set us up for calm immune systems and optimal neurotransmitter production.

In general, when you're craving carbs, don't beat yourself up or deprive yourself, but also don't just reach for the processed version of those foods. Reach for starchy veggies. They'll hit the spot and give your body the fuel and benefits of carbs without the inflammation.

I'll also acknowledge that there are those who do better on a keto or low-carb diet of mainly protein, fat, and vegetables. I see these diets working best for my patients with significant insulin resistance, bipolar disorder, or seizure disorders and for male bio-hackers committed to optimizing their physiology. On the other hand, for women of reproductive age, in my practice, I have seen low-carb diets play out in a few different ways. For those who commit to it wholeheartedly and eat sufficient nutrients, they seem to retrain their bodies to rely on alternative fuel sources, and they thrive. But for women of reproductive age who drop in and out of low-carb and keto diets or fail to meet their caloric needs, I see their bodies buckle under the unpredictable supply of carbohydrates, and develop symptoms such as irregular periods, fatigue, and an exacerbation of insomnia and anxiety. I believe this occurs because, during the reproductive years, a woman's body is constantly surveying the environment and assessing whether there is abundant food availability, as well as enough fat stores, to undertake a pregnancy. If there is, a woman's hormones continue to cycle normally, she ovulates every month, and the downstream hormones involved in her hypothalamic-pituitary-adrenal (HPA) axis can function properly as well, helping her to maintain a positive, motivated, and calm mood. However, if her body gets the signal that there is food scarcity—which can happen when avoiding carbohydrates—then the HPA axis reaches the conclusion that it would not be an optimal time to conceive, and it can adjust the hormonal cascade in a way that prevents the woman from ovulating. This has downstream ramifications on the rest of the body—and can lead to anxiety.

YOUR BRAIN IS MADE OF FAT

Finally, let's talk about fat. I do *not* recommend that anyone struggling with anxiety eat a low-fat diet. In fact, one of the fastest ways to improve anxiety is to *increase* the healthy fats in your diet.

When it comes to fat, I want you to forget what you've been taught about saturated and unsaturated fat. Focus instead on the distinction between naturally occurring fat and man-made fats, such as trans fats, margarine, and industrially processed vegetable and seed oils—for example, canola and soybean oil. You might even be under the impression that vegetable oils are good for you—after all "vegetable" appears in the name. Unfortunately, while olive oil and avocado oil are perfectly healthy, the industrial vegetable and seed oils are highly processed and inflammatory,[40,41] and they can promote heart disease, cancer, and other health issues.[42,43,44,45,46] Meanwhile, the fats from animals and minimally processed fat from plants, such as avocados, nuts, and coconut, are generally well tolerated by the body and promote health, even when they're technically "saturated" fats, as is the case with coconut and macadamia nuts. These sources of fat are more closely related to the fats humans have been consuming for millennia and so are more recognizable to the body and less likely to provoke the immune system. Once you've ditched canola oil, you may be wondering what fats to cook with instead. For low-heat cooking, use olive oil, coconut oil, and grass-fed butter; for high-heat cooking, ghee, beef tallow from pasture-raised animals, and avocado oil are all good choices. Even if you're not cooking with canola oil at home, it's worth being cognizant of the fact that anytime you go out and exchange money for food, that food is almost certainly being prepared with industrially produced vegetable oils. As such, eating home-cooked food can help prevent a lot of inflammation.

I generally recommend that my patients with anxiety increase their consumption of healthy fats. Of course, moderation is important when it comes to fat, but unlike refined carbs, sugar, and processed foods, real fat is satiating and thus harder to over-consume. Eating a reasonable amount of fat with every meal helps to stabilize blood sugar and prevents your body from experiencing the shakiness and irritability of false anxiety.

SNACKS, FRUIT, AND FERMENTED FOODS

Now that we've built your plate, let's cover snacks.

First: fruit. Here's a hot take: Eat fruit and don't overthink it. Some folks are concerned about the sugar content of fruit. Of course, be reasonable about it—use caution around sugar bombs such as fruit smoothies and dried fruit. But if you want to have an apple or some berries at the end of your meal, go for it, and feel good about it.

In addition to fruit, snack on nuts and seeds, olives, avocados, hard-boiled eggs, grass-fed jerky, and good-quality dark chocolate (aim for high cacao content, minimal sugar and dairy, and no genetically modified soy lecithin). You can also snack on sea vegetables such as dulse and nori, which are an excellent source of hard-to-find nutrients such as sulfur and naturally occurring iodine.

I also recommend building fermented foods into your day—this includes kimchi, sauerkraut, miso, natto, apple cider vinegar, beet kvass, and, if you tolerate dairy, perhaps some kefir and yogurt. These are functional foods—by introducing beneficial bacteria to the gut, they will help heal your gut, decrease inflammation, promote the synthesis of neurotransmitters such as serotonin and GABA,[47,48,49] and improve your immune function as you snack.

A final food to embrace is bone broth. This source of collagen, glycine, glutamine, and iron is a nutritional workhorse that heals your gut and builds up your nutrient stores, helping your skin, hair, and nails stay healthy—all of which are markers of the health of your internal landscape as well. Most traditional cultures had their version of bone broth because they understood it to be a nutritional necessity and an efficient way to derive as much nutrition as possible from the animals they consumed.[50]

The Best of Both Worlds

Many of us would do well to draw on the best of veganism and the paleo diet. At first glance, these approaches seem diametrically opposed—a face-off between the yoga girl drinking green juice and the caveman CrossFitter gnawing on a bone, stereotypically speaking. But combining the benefits of these two disciplines offers a useful both/and approach to eating.

My vegan patients typically eat plenty of fruits and vegetables, which is great, but many of them are also drawn to sweets (like heavily processed vegan cupcakes) and excessive quantities of nut butter. When my vegan patients struggle with anxiety, I think it's often related to both micronutrient deficiencies (e.g., zinc, B_{12},[51] and omega-3 fatty acids) and too much "cold" food—that is, in the Chinese medicine sense, the vegan diet is lacking in grounding, warming, substantial foods such as bone broth, chicken soup, and beef stew.

On the flip side, we have Joe Paleo, who rewards himself after a long day in the CrossFit "box" with a large hunk of pork belly for dinner. Paleo? Check! Too much meat and not enough vegetables? Also check! This is a common pitfall.

At the end of the day, most people following a vegan diet might feel (physically) better if they included a little bit of bone broth and red meat. And most people eating a paleo diet could stand to eat a little less meat and a few more greens. Eating a plant-strong diet that includes judicious amounts of animal foods, such as unprocessed meat and seafood, is generally a good bet for optimal physical and mental health.

WHAT *NOT* TO EAT

Certain foods can trigger inflammation, harm the gut, cause blood sugar spikes, and otherwise negatively affect your body and brain. When it comes to eating for better physical and mental health, these are the most important foods to avoid or minimize:

- Highly processed foods.
- Man-made fats, which include trans fats, margarine, and industrially processed vegetable and seed oils, such as canola oil.
- Added sugar and high-fructose corn syrup.
- Artificial sweeteners.
- Refined carbohydrates. (But take care especially to avoid conventional American flour, which comes from crops sprayed with Roundup, a pesticide that contributes to gut inflammation,[52] intestinal permeability,[53] and cancer risk,[54] yet somehow is still in our food.)
- Food from other genetically modified crops, such as conventional soybeans and corn. (Controversial as this may be, *my* gut says that GMO-associated pesticides are damaging our guts and therefore adversely affecting the functioning of our immune systems and brains.)
- Anything with preservatives, food coloring, or the ingredient "natural flavor," which is often anything but natural.

That's a pretty long list. Can or should you follow it perfectly 100 percent of the time? *No.* Do I? *Nope!* The world we live in can make it hard to eat this way; it's expensive, inconvenient, and socially isolating. And the truth is, the stress of trying to achieve perfection with food will contribute to your anxiety more than optimal healthy eating will help relieve it. Simply do your best and be flexible with this process *and with yourself.*

Count Chemicals, Not Calories

You might have noticed that calories are not my focus. When it comes to your body's health, I believe it's the chemicals that matter. (And yes, everything, even water, is technically a chemical, but I use the term "chemical" to signify man-made food components, such as preservatives and artificial sugar.) You will still hear conventional folks preaching that weight loss comes down to a simple equation of calories in minus calories out, but this formula is based on bad science. The *quality* of the calories we eat matters more for health than the quantity. In fact, the *quality* of the food we eat impacts the *quantity* that we feel driven to eat—that is, certain calories are satiating while others seem to only whet the appetite. Not to mention that the *quantity* of food we eat impacts our basal metabolic rate, which is the "calories out" portion of the equation. This is calculus, not subtraction.

For any remaining calories-in-calories-out stalwarts, allow me to disabuse you of this notion with some evidence. Researchers from Canada fed two populations of mice the same diet and same number of calories per day—but one group received water sweetened with aspartame, while the other group received unsweetened water. Though there was no difference in *calories* between the two groups, the aspartame-consuming mice gained weight and developed markers of metabolic syndrome (high cholesterol and dysglycemia).[55] The mere presence of artificial sweetener was sufficient to derange the animals' metabolic health.[56]

A recent study on humans also elucidated this concept. Researchers in Sweden gave two groups of nonobese healthy adults the same number of calories per day, with the only distinction being that one group snacked on peanuts while the other group snacked on a calorically equivalent amount of candy. While the peanut snackers saw improvements to their metabolic health, the candy-consuming group experienced an increase in weight, waist circumference, and LDL cholesterol, a result that illustrates that *what* we eat, not just *how much*, impacts our metabolic health.[57]

Processed foods, which are engineered to be hyperpalatable, override our satiety signals, driving us to overeat and arrive at a state of poor metabolic health, which negatively impacts brain health. Your best shot at reducing this source of false anxiety is to aim to eat real food and avoid fake food—focusing more on the "what" than on the "how much."

Chapter 8

Body on Fire

We are, in the end, a sum of our parts,
and when the body fails, all the virtues
we hold dear go with it.

—*Susannah Cahalan,* Brain on Fire

Since the 1990s the monoamine hypothesis has been psychiatry's prevailing theory of certain mental illnesses. This theory suggests that mental health issues result from a genetic imbalance of neurotransmitters such as serotonin in the central nervous system. Today the monoamine hypothesis remains the consensus for how we make sense of conditions such as depression and anxiety. There is certainly merit to this theory, but there is now arguably stronger evidence to support the idea that often—not always, but *often*—inflammation in the body plays a central role in depression and anxiety.[1,2,3] This competing theory is called the cytokine, or inflammatory, hypothesis.

The broad idea is that our evolutionary response to inflammation lines up with symptoms of mood disorders. Most of the symptoms we associate with feeling sick—fatigue, malaise, *ugh*—are caused not by the virus or bacterium itself but by our own immune system mobilizing to fight it. And acute inflammation—which is essentially the immune system in action, ramping up to protect the body—is a battle that occurs in our bloodstream. By evolutionary design, these symptoms make us cancel plans, get under the covers, and rest; this approach to healing prevents us from infecting others and gives us our best shot at fighting the invader, as the immune system functions best when we're at rest. But those symptoms bear an eerie resemblance to what we call depression.

Further bolstering this theory that inflammation can affect mental health is recent research showing that cytokines—signaling proteins that are secreted by specific cells of the immune system and essentially serve as markers of inflammation—can cross the blood-brain barrier, meant to protect against circulating toxins and pathogens. In fact, it has been shown that cytokines directly impact regions of the brain involved in fear and threat detection—including the amygdala, insula, medial prefrontal cortex, and anterior cingulate cortex. This suggests that inflammation can directly contribute to anxiety by informing the brain that we are indeed under threat.[4] Consequently, cognitive symptoms that are commonly regarded as mental health issues—such as anxiety and the intrusive thoughts of OCD[5,6]—may in fact stem from the brain's response to inflammation.

THE ARMY WITHIN

The human immune system is an exquisitely designed, complex network of cells and signaling molecules that serves as a powerful defense system against an array of threats. Since the earliest humans walked the earth, the sophisticated weaponry of the immune system has been a lifesaver in the face of bacteria, viruses, and other pathogens. However, changes to our environment have impacted the functioning of our immune systems. For starters, advances in medicine and modern hygiene—such as antibiotics and water sanitation—have allowed our immune systems to take benchwarmer status (with the glaring exception of the recent COVID-19 pandemic). Our bodies no longer need to spend as much time fighting big-ticket invaders, which means they receive less "education" in detecting friend versus foe. At the same time, our bodies are increasingly bombarded with unrecognizable chemicals and foods—ranging from pesticides to phthalates to Pop-Tarts (essentially, foreign agents our bodies didn't evolve to deal with)—that provoke the immune system in much the same way a genuine infection would. A daily ingestion of Doritos, for instance, leaves the immune system belligerent and confused. It keeps fighting, thinking it stands a chance at killing off the "infection" of Doritos, but our immune system isn't built to defeat chips—not to mention that we get "reinfected" with every snack. Over time, a consistently inflammatory diet can result in a dysregulated, hypervigilant immune system, an inflamed body, and sustained feelings of depression or anxiety.

WHEN THE BODY TURNS ON ITSELF

Anytime I watch network television, I'm struck by the number of ads I see for immunosuppressive drugs—medications to treat a misfiring immune system. Autoimmune conditions, including rheumatoid arthritis, ulcerative colitis, Crohn's disease, celiac disease, Hashimoto's thyroiditis, Graves' disease, multiple sclerosis, lupus, psoriasis, vitiligo, type 1 diabetes, and eczema, have reached epidemic levels in the United States, affecting an estimated ten million to forty million Americans[7,8]—and the number increases each year.[9]

Autoimmunity occurs when a chronically misinformed immune system begins to attack the body's own cells. The preconditions necessary to develop autoimmunity seem to include a stressor (this can be physical stress, such as an aggressive bout of food poisoning, or mental stress, such as experiencing a sudden loss); a genetic predisposition; and a compromised intestinal barrier,[10] also known as intestinal permeability or leaky gut. Damage to our tissues is also hypothesized to play a role in provoking immune attack, which can occur as a result of chronic infection, injury, or environmental exposures such as to mold or heavy metals.[11,12] Symptoms of autoimmunity might show up as rashes, digestive issues, joint pain, or fatigue, or in a host of other ways. But do you know what other symptoms track closely with autoimmune disease? Depression and anxiety. In fact, they occur at higher rates than with other serious illnesses, which suggests that people aren't just depressed and anxious as a result of being sick,[13] but that these mood changes are a direct result of something inherent to autoimmunity.

My theory is that the chronic immune provocation itself can cause depression and anxiety because cytokines, the inflammatory markers that directly impact the brain when we're inflamed, essentially convey to the central nervous system that the body is in a state of battle, causing us to feel uneasy.

In fact, the connection between anxiety and autoimmunity seems to be bidirectional. Research on adverse childhood experiences (referred to in studies as ACEs) has found significantly higher incidences of autoimmunity among people who were neglected, abused, or otherwise subjected to serious stressors in their youth.[14] I sometimes wonder if certain cases of autoimmunity develop later in life as a physical response to the lifelong pain inflicted by childhood trauma. Perhaps adults who have experienced ACEs have bodies that have internalized pain—and have been conditioned to have a hypervigilant response to threat, both mentally and physically.

On a hopeful note, autoimmunity may be preventable and remediable if we take steps to ameliorate the contributing GI issues, environmental exposures, and stress. My patient Nina suffered from chronic irritable bowel syndrome (IBS), mood changes, a butterfly-shaped rash on her face, and frequent oral canker sores—all characteristics of a common autoimmune condition called celiac disease, caused in susceptible individuals by eating gluten. Nina was also noticing GI distress when she would eat gluten-rich meals such as pasta. When Nina consulted with a GI doc, he presented her with negative test results and exclaimed, "Good news, you don't have celiac!" Meanwhile, he offered *no explanation* for why gluten was making her double over in pain. His advice: *Soldier on! You can, and should, continue to eat gluten.* Though conventional medicine failed her, Nina managed to connect the dots between her chronic digestive issues, her mysterious autoimmune-like symptoms, and the gluten in her diet. Within days of going gluten-free, her deteriorating health changed course, and her body started to mend.

Unfortunately, conventional medicine, with its emphasis on treatment rather than prevention as well as its view of the body as composed of several separate organs rather than a web of interconnections, is not well positioned to address autoimmunity. One doctor treats your psoriasis while another treats your anxiety. Who is stepping back to see the relationship between the two? Who knows how to soothe the inflammation at the root of both, rather than simply suppressing it with steroids and immune-modulating medications (which cause a rebound of inflammation upon discontinuation)? Conventional medicine also prefers to wait until an illness is "full blown" before treating it, which not only causes needless suffering but also makes it harder to resolve the problem once it is finally addressed.

If you suspect autoimmunity plays a role in your anxiety—or if you have any signs or symptoms similar to those I have mentioned—see a functional medicine doctor or naturopath who can get to the root of your condition. I also recommend following the advice outlined in this book for healing the gut, decreasing inflammation, and managing stress.

THE GUT-IMMUNE-BRAIN CONNECTION

I have a patient, Joni, forty-two, who had, in addition to anxiety, many telltale signs that something was awry in her digestive tract, as evidenced both by her near-constant state of bloating and abdominal distension and by bowel habits that alternated between constipation and loose stool. She was also experiencing acne, eczema, and migraines, which are additional clues that lead to the gut. We spent a few months taking a concerted approach to healing her gut and, right on cue, as her digestive symptoms and inflammatory symptoms improved, so, too, did her anxiety. Perhaps even more supportive of this connection: anytime Joni would accidentally—or intentionally, for that matter—consume a food that inflamed her gut, her anxiety would return precipitously.

As we now know, approximately 70 percent of our immune activity is stationed in the gut.[15] This distribution may seem surprising, but it's actually a good design, given that the gut is one of the body's primary contact points with the outside world. The small intestine alone has the surface area of a tennis court, granting the food and drink and germs that we ingest direct access to our bodies, and making the digestive tract the logical place to station much of our defensive army. We have skin surrounding our outsides to protect us, but everything we swallow gains intimate access to our internal landscape.

Why does gut health affect anxiety? There are a few main reasons. First, your brain and your gut are talking to each other, even if your psychiatrist and gastroenterologist are not. As I described in chapter 2, there's a direct hotline between the gut and the brain. This communication is made possible by the vagus nerve, which is the principal component of the parasympathetic nervous system, and it sends information in both directions between the gut and the brain.[16] So, your brain might say to your gut: *I'm about to speak in front of a hundred people, and I'm freaking out. Let's have diarrhea!* But, perhaps more importantly, your gut also talks back to the brain, with a remark like: *Ever since that antibiotic you took for that UTI, everything's been a mess down here. I'm going to make you feel anxious and unwell so you will rest until we get this situation sorted out.* When you ignore these signals, your gut never has an opportunity to repair itself, and it will keep sending the message to your brain to make you feel lousy.

In addition to vagus nerve communications, the trillions of microorganisms in our digestive tract, referred to as our microbiota, also impact our anxiety levels. There is an emerging field called psychobiotics that studies the effects of GI microbes on mental health and has begun to establish a bidirectional relationship between gut microbiota and the brain, termed the microbiota-gut-brain (MGB) axis. Mounting evidence, as described by psychobiotics researchers at Tufts University, points to the possibility that gut microbiota can influence the immune and nervous systems and vice versa.[17] Researchers have also uncovered that certain species of *Lactobacillus*, for example, help improve our stress resilience and cognitive symptoms, and alleviate anxiety.[18] Other studies have focused on the relationship between mental health and antibiotics—which kill off pathogenic and beneficial bacteria alike—revealing that "recurrent antibiotic exposure is associated with increased risk for depression and anxiety."[19] Antibiotics are thought to impact mental health by destroying the populations of beneficial bacteria in the gut. Indeed, several studies have demonstrated that the restoration of healthy beneficial bacteria through probiotic supplementation or the consumption of fermented foods has the ability to improve mood and anxiety.[20,21] Last, research has shown that reinstating beneficial bacteria—which can be done through diet and probiotic supplementation—reduces systemic inflammation, which is thought to be one way that beneficial bacteria exert their antianxiety effect.[22] While this is still a nascent field, it is already overwhelmingly clear that improving the balance of microbes can go a long way toward reducing anxiety.[23]

You'll recall, too, that the gut, along with the brain, is responsible for producing certain neurotransmitters, including serotonin and GABA—chemicals our brains need to help us feel good. As I've mentioned, there is also evidence in the medical literature suggesting that specific *Bacteroides* strains of gut bacteria are involved in the synthesis of GABA.[24] The fact that certain bacteria are involved with GABA synthesis helps us understand how a compromised gut flora can directly impact GABA availability and, therefore, anxiety. So, when I survey the anxiety epidemic before us, *my* gut tells me that the antibiotic-induced decimation of GABA-producing bacteria in our guts plays a central role.

Finally, a particularly important way that gut health contributes to anxiety is via inflammation. As I explored in Part I, gut dysfunction has a unique ability to create systemic inflammation throughout the body, including the brain. Put simply: fire in the gut, fire in the brain. And as we discussed, this can occur either when lipopolysaccharides—a normal constituent of the gut—seep through a leaky gut wall, or when the gut is overrun by pathogenic and opportunistic bacteria and it instructs the immune system to go into a state of red alert and be combative rather than tolerant.[25] In both these scenarios, the inflammation reaches our brain, and we feel anxious.

A quick technique for identifying whether your gut is playing a role in your anxiety—beyond observing GI symptoms such as diarrhea, constipation, bloating, and heartburn—is to sketch out a timeline of your physical health, life events, and anxiety. For some of us, anxiety has been a constant state of mind for as long as we can remember. For others, anxiety started during a divorce, for example, or after a trip abroad, a course of antibiotics, a medical procedure, or a bout of food poisoning. If any of these latter types of explanations are the case, there's reason to suspect that at the heart of your anxiety is compromised gut health, likely brought about by changes to the microbiome, either from the invasion of a microbe or from the decimation of your beneficial bacteria following antibiotic treatment. Additionally, significant stress can itself behave much like a course of antibiotics, impacting the ecosystem in your gut and the health of your gut lining. So, regardless of whether the original cause of the problem was Cipro, salmonella, or a divorce, once you heal the gut, you'll have gone a long way toward healing the anxiety.

Gut healing comes down to a few steps: removing what irritates the gut, adding in what soothes the gut, and then creating the conditions for the gut to heal. As we've discussed, various foods can irritate the gut. While gluten and dairy get a lot of airtime,[26] industrial vegetable oils,[27,28,29] artificial sweeteners,[30] food stabilizers such as carrageenan,[31] and pesticides such as Roundup[32] are hidden causes of gut dysfunction in our diets. There are also several classes of medications that adversely affect gut health. You will best protect your gut if you can err on the side of avoiding unnecessary antacids,[33] painkillers such as ibuprofen[34] and prescription pain meds,[35] birth control pills,[36] and antibiotics[37] whenever possible. Of course, these are broad suggestions that need to be taken in the context of your unique health needs (don't make any changes without a conversation with your health care provider). And I don't mean to suggest that you should avoid all of these 100 percent of the time. This is just to suggest weighing the benefits of the medication with the costs to your gut health.

When it comes to soothing the gut, I recommend bone broth to provide the collagenous matrix to heal a damaged gut lining; ghee, which provides butyrate, a natural fuel source for the cells of the gut lining; and glutamine, an amino acid that helps repair these tissues. You'll also want to repopulate your gut with beneficial bacteria. Most people think of probiotics—healthy, live microorganisms found in pill form—for this. While probiotics can be helpful, the real cornerstone of colonizing your gut with healthy bacteria is to regularly consume fermented foods, such as sauerkraut, along with starchy tubers, such as sweet potatoes (which function as the prebiotic, or food, for the beneficial bacteria to eat). It's worth noting that some folks need to take extra care around introducing beneficial bacteria. If you have symptoms such as bloating, gas, or burping, as my patient Joni did, then you might have an overgrowth of bacteria in your upper GI tract, called small intestinal bacterial overgrowth, or SIBO. This would mean it's preferable to kill off some of that overgrowth before reintroducing good bacteria. In this scenario, you would want to work with a naturopath or functional medicine practitioner to manage your SIBO.

The last piece of gut-healing advice I have: get out of your own way. Gut healing requires an allocation of energy, and this can only be achieved through rest. As a culture, we love to buy the latest supplement to fix a problem, but the necessary behavioral changes can sometimes feel like an afterthought. However, after years of working with patients on gut healing, I've come to recognize that this one is a sine qua non. In other words, if you don't rest and reduce stress, the gut won't heal, no matter how gluten-free and bone-broth-full you become. If you have gut issues and anxiety and want to feel better, your body needs rest. I know it can sometimes feel as if you need to move mountains to actually fit in adequate sleep and stress management. But move those mountains. Get enough sleep, do less overall, and get in the habit of practicing some form of relaxation every day.

Note to anyone taking a stimulant medication, such as Adderall or Vyvanse: Healing in general and gut healing in particular can occur only when the nervous system is in a parasympathetic or relaxation tone. Be aware that stimulants are putting your body into a stress response *by design*. I've had many patients who were unable to heal their guts until they discontinued their stimulant use, because their bodies were in a stimulant-induced stress response, which was standing in the way of true healing.

The final step to gut healing is to get a squatty potty. Google it—and thank me later.

HOW TO TREAT ANXIETY BY
RECALIBRATING THE IMMUNE SYSTEM

It's possible to use food as information to restore the immune system to a state of balance and help bring depression and anxiety back under control. In addition to decreasing consumption of inflammatory foods, I work with my patients on adding in certain foods and practices that will help extinguish inflammation. I like to start with curcumin, the active compound in turmeric. Curcumin acts on a particular fulcrum of the immune system called NF-κB[38,39]—essentially communicating to it, *Don't sweat the small stuff.* I recommend patients incorporate curcumin into their diets with curry dishes or by blending turmeric paste with ghee and water to make a traditional Indian "golden milk." Cooking with turmeric alongside black pepper has a synergistic effect, offering even more anti-inflammatory value. Ginger,[40] garlic,[41] onions,[42] omega-3–rich foods such as salmon,[43] and nearly every pigmented vegetable,[44] such as dark leafy greens and beets, are also beneficial to an irascible immune system.

Beyond decreasing inflammation, it's also necessary to *recalibrate* the immune system. This comes back to the notion that the immune system is powerful and complex machinery that evolved under vastly different conditions than those we live under today. In what I'll call "evolutionary times," we lived in a state of constant exposure to microbes. Some of these microbes were benign, some symbiotic (meaning they helped us), and some pathogenic and parasitic (meaning they had the potential to harm us). From the birth canal to the dirt on our food to our proximity to animals (as well as their feces), we were exposed to microbes. And all these exposures created a diverse ecosystem of microbes living in our digestive tracts. This ecosystem of bugs *calibrates* the immune system—that is, the bacteria and viruses in our body are in constant conversation with the immune system. With a diverse ecosystem, and therefore a wide-ranging conversation, the immune system gets a lot of practice, and it learns to recognize the difference between pathogenic and benign, between friend and foe.

Cut to today. From the first moment of life, many of us start off with compromised gut flora. For instance, babies delivered by cesarean section—which currently make up approximately 32 percent of American births[45]—have their digestive tracts imprinted with microbes from hospital air and the skin of hospital employees, rather than Mom's vaginal flora.[46] A vaginal birth imprints the baby's digestive tract with a diverse array of microbes from the mother, but even this transfer of beneficial bacteria can be compromised by the broad-spectrum antibiotics we give to mothers positive for group B strep—a normal vaginal bacterium—during labor. (In the United States, about 40 percent of women receive antibiotics during delivery.[47]) In no way do I intend to blame or shame anyone for having a C-section—which is often an unavoidable and lifesaving procedure—but these are, nonetheless, common aspects of modern life that impact our microbiota and contribute to our dysregulated immune systems. And we should be clear-eyed in our efforts to adjust accordingly. "I do not wish to ban antibiotics or Cesarean sections any more than anyone would suggest banning automobiles," writes renowned microbiome researcher Dr. Martin J. Blaser in his book *Missing Microbes*. "I ask only that they be used more wisely and that antidotes to their worst side effects be developed."[48]

Even after we're born, we move through a world that makes a broad assault on the diversity of our gut flora. By age twenty, for instance, the average American young person has received seventeen courses of antibiotics.[49] Once you add to that the sugar and processed foods in our diets, chronic stress, alcohol, antacids, birth control, psych meds, antibiotic residue in our dairy products, and chlorine in our tap water, the ecosystems of our modern digestive tracts are decimated. In the end, we're left with a commanding immune system that has essentially not gone through basic training. This is evident in the chronic inflammatory conditions that are epidemic among children today—asthma, eczema, allergies, and food intolerances. I would even argue that some cases of attention deficit hyperactivity disorder (ADHD) are caused by inflammation.[50] And this is in addition to the lesser-known connection between chronic inflammation and many of the most significant health issues plaguing our population, from heart disease[51] and cancer[52] to dementia[53] and, of course, mental health issues.[54,55]

So, how does one go about addressing this problem? Some factors are not entirely in our control, but there *are* certain steps we can take. Below is a list of the recommendations I offer my patients—many of which conveniently overlap with the advice I offer for reducing anxiety in general.

1. Avoid inflammatory foods, such as vegetable oils, and add anti-inflammatory foods, such as turmeric.
2. Maintain a diverse ecosystem of microbes in your gut by avoiding unnecessary antibiotics and consuming fermented foods as well as starchy tubers.
3. Heal the gut by avoiding what irritates it, such as foods exposed to the herbicide Roundup (e.g., conventional flour), and adding in what soothes it, such as bone broth.
4. Give your immune system the nutrients it requires to function

properly—zinc, vitamin C, vitamin A, and the B vitamins are a good start.

5. Avoid blue light after sunset. Immune activity is promoted by the hormone melatonin, but we only secrete melatonin in response to darkness. (See chapter 5 for more information on blue light.)

6. Get enough rest. Our immune system does most of its work while we sleep.

7. Relax. When we exist in a state of chronic stress, immune activity is suppressed.

8. Get your vitamin D level checked, and be sure you receive healthy levels of sun exposure (see the next section for more information). If getting sufficient sun exposure is not possible, supplement with vitamin D.

As usual, modern life is basically the opposite of these recommendations. Our food leaves us inflamed and undernourished; we don't get enough rest; we're indoors far too much—lacking full-spectrum sunlight during the day yet over-illuminating our environments at night; and we exist in a fear state, addicted to the twenty-four-hour news cycle. But if you can make even some of the changes from the list above, you'll be well on your way toward a better-calibrated immune system and a less anxious mind.

WE GOT SUNSHINE WRONG

Sunshine boosts our mood in a variety of ways, but perhaps its most significant contribution to our mental health is that it allows our skin to manufacture vitamin D. All hail vitamin D, which is so much more critical to our health than most of us appreciate. In order to have a calm mood and avoid colds and autoimmunity, we need healthy vitamin D levels.

Vitamin D is, indeed, so essential that evolution took no chances. Our bodies are equipped with machinery that allows us to manufacture vitamin D in response to something so reliable and ubiquitous, we will never be without it—the sun. This was a brilliant design—until, that is, sunblock, video games, and working from home basically eliminated sunshine from our lives.

So, what's the big deal about vitamin D, and why would evolution make it such a priority? It's almost a misnomer to call it a vitamin, as it functions more like a hormone in the body. Vitamin D is critical for proper immune function (thereby helping to rein in infections[56] and, arguably, nascent cancers as well[57]), and, as I mentioned, it calibrates the immune system, allowing us to mount an attack on genuine pathogens (e.g., COVID-19),[58] while not overreacting to benign allergens (e.g., asthma, seasonal allergies)[59] or pointing the machinery of the immune system at the wrong target (i.e., autoimmunity).[60] Healthy vitamin D levels are relevant for cognitive function and dementia prevention,[61] cardiovascular health,[62] bone density and osteoporosis prevention,[63] fertility and hormone health,[64] and the prevention of certain cancers.[65,66,67,68,69,70] Vitamin D is also critical for mental health; healthy vitamin D levels are, as research has shown, associated with lower rates of depression and anxiety.[71,72]

The World Health Organization's party line is that five to fifteen minutes of sun per day two to three days per week is adequate to make sufficient vitamin D in our bodies. In reality, however, those fifteen minutes aren't cutting it. Studies show that a majority of Americans have insufficient vitamin D,[73,74,75] and that's going by the established lab range of 30 ng/mL, which is likely set too low (meaning, if the goal were optimal vitamin D levels closer to about 50 ng/mL,[76] even more of us would be considered deficient). I've checked enough bloodwork in my practice to be convinced that we're all walking around with empty vitamin D tanks. When in doubt, I like to refer back to the conditions in which we evolved. Let's ask ourselves, *Were we outside for fifteen minutes two to three days a week?*

We've been taught to fear the sun and slather sunblock across every inch of our bodies. And there's no question that avoiding overexposure to the sun prevents some cases of skin cancer and a number of deaths. But I wonder if this high degree of caution around sun exposure has also come at the cost of a population-wide vitamin D deficiency, impacting folks with melanated skin most profoundly.[77] It's time to reshape the conversation around the risks and benefits of the sun, understanding that, while the risk of skin cancer is real, perhaps we have swung too far in the other direction.

Our relationship to the sun poses a delicate balance between the need for vitamin D and the risk of skin cancer. Skin color is the way our bodies resolve these competing priorities. If our ancestors evolved somewhere with ample sunshine, then the melanin, or pigment, sits close to the surface of the skin cells, making the skin appear darker and providing a strong protective shield against DNA damage from the sun's rays and thereby skin cancer. Conversely, if we descended from people living in places with scant sunshine, then the melanin sits deeper in the cells, affording little protection against skin cancer but making it easier to manufacture vitamin D from less sunshine. The vitamin D–skin cancer balance in the body is so crucial that genes for skin color evolve rapidly in response to changing conditions. These determinants of skin color, such as the SLC24A5 and MFSD12 genes, are considered a "highly conserved" area of the genome and a strong target of natural selection.[78] There was even a "selective sweep" of these genes, in which their expression changed in all Eurasians about seventy thousand years ago, around the time that these populations migrated out of Africa.[79] In other words, evolution doesn't mess around with this equilibrium because there's not a lot of wiggle room in either direction: skin cancer can be lethal, but insufficient vitamin D compromises health across the body.

When it comes to skin cancer prevention, the most important thing is to avoid getting a sunburn. Ironically, though, trying to entirely avoid exposure to the sun out of fear of skin cancer can actually leave us more vulnerable on those rare occasions when we do finally get some sun. Many of us keep ourselves as pale and protected as possible, but when we happen to take a sunny vacation or find ourselves out of sunblock while at a parade in June, we're then at higher risk of getting burned. This sunburn then puts many of us at *greater* risk for skin cancer than if we'd simply been getting baseline low levels of sun exposure year-round.

Meanwhile, the one-size-fits-all advice to avoid sun exposure—no matter its well-meaning intent—also does a disservice to those with darker skin. If you have melanated skin, it's very unlikely that fifteen minutes of daily North American sunshine will allow your body to manufacture sufficient vitamin D. And what the medical establishment isn't telling you is that you have a relatively low risk of developing sun-associated skin cancer.[80,81,82] However, the consequent vitamin D deficiency from avoiding the sun *is* associated with diabetes,[83,84] obesity,[85] risk of death or ICU admission from COVID-19,[86,87] dementia,[88] cancer in general,[89,90] heart disease,[91,92,93] osteoporosis,[94] asthma,[95] autoimmunity,[96,97] depression,[98] and anxiety.[99]

In reality, the right amount of sun exposure is as varied as the combinations of skin tone and latitude are. We all have a different risk-benefit analysis to make, based on our skin tone, our family history of skin cancer, our place of residence, our lifestyle, and even our personal history of sunburns in childhood. It's also helpful for us to consider how our ancestors' geography lines up with where we currently call home. An African American person in Chicago may be more at risk for vitamin D deficiency than for skin cancer, while a person of Northern European descent living on the equator should pay more mind to their risk of melanoma than the downstream effects of vitamin D deficiency.

While it may seem that getting your vitamin D in pill form saves you from needing to expose your skin to the so-called harmful rays of the sun, it is my conviction that supplements don't take the place of the real thing. As with so many aspects of human health, vitamin D dynamics are more complicated than what a pill can achieve.[100,101] Studies have repeatedly shown that low vitamin D levels are associated with various disease states and bad outcomes—but research then fails to demonstrate that restoring healthy levels of vitamin D with supplementation does anything to correct the problem. The missing link between these two findings may be that there is more involved in the body's response to the sun than what is reflected in a simple vitamin D level. Indeed, humans make several other important "photoproducts" in response to UVB light,[102,103] such as beta-endorphin, adrenocorticotropic hormone, calcitonin gene-related peptide, nitric oxide, and substance P, which can help with pain, stress, high blood pressure, and inflammation.[104,105,106,107,108,109]

Sun exposure, in particular, seems to be beneficial to our health and well-being, and research even suggests that these benefits may offset the increased mortality risk from skin cancer.[110] A recent twenty-year study following nearly thirty thousand subjects in Sweden found that individuals avoiding sun exposure were twice as likely to die from "all cause mortality," meaning people who avoid the sun are twice as likely to die from all causes of death.[111] So, while responsible supplementation (balanced with vitamin D testing and supplementation of other relevant micronutrients such as magnesium, phosphorus, and vitamins A and K) is a wonderful strategy to round out our vitamin D needs, especially in the winter, when it comes to overall well-being, there are specific advantages to sun exposure that are difficult to measure and go well beyond raising vitamin D levels.

Lyme Disease and Mold

Going into the intricacies of Lyme disease and a condition called Chronic Inflammatory Response Syndrome (CIRS)—often precipitated by mold exposure—is beyond the scope of this book. And yet I would be remiss if I did not at least point to the fact that, in many cases, anxiety is a symptom of the often pronounced state of immune activation that can be a consequence of either of these illnesses. If you have reason to believe that you might have contracted Lyme disease from a tick bite, or if you have been living or working in a building with water damage or mold, it's worth seeking a consultation with a functional medicine practitioner to get evaluated and treated for these conditions. Both Lyme and mold can cause profound degrees of anxiety, and they are frequently overlooked or dismissed in the hallowed halls of conventional medicine.

GLUTEN AND DAIRY, A CLOSER LOOK

There are two reactions you might have upon reading the words "gluten and dairy": you're rolling your eyes because you're sick of hearing about this made-up sensitivity that everybody keeps going on about, or you're rolling your eyes because this conversation is so played out—you've read every book and heard every wellness podcast, and you could write your own thesis about gluten and dairy. Well, here's some new information, seen through the lens of anxiety, that is well worth a second look.

I get it: gluten intolerance is everybody's favorite dietary intolerance to mock—and cheese *is* divine. It's no wonder we get so defensive when someone tries to take away our gluten and dairy—this couplet makes up our most popular socially condoned drug. Notice how every American comfort food is a combination of gluten and dairy: milk and cookies, macaroni and cheese, grilled cheese sandwiches, ice cream cones, brownies à la mode, and pizza. These ingredients are associated with nearly every culinary comfort and joy this world has to offer. So, I wouldn't blame you for thinking, *Why would this sadistic doctor want to take away these basic human rights?*

With my condolences, I'm going to insist that we talk about the direct role gluten and dairy may be playing in your anxiety. By causing gut irritation, systemic bodily inflammation, and the cycle of addiction, these foods dial up our inflammation levels and behave like any other drug, lifting us temporarily only to drop us into withdrawal soon after and leave us strung out and anxious.

Today we are suddenly seeing sweeping levels of nonceliac gluten sensitivity,[112] which is odd, especially considering how we've been eating gluten for thousands of years without too much difficulty. But I can't deny the fact that so many of my patients with anxiety feel better when they get off gluten. Here's the mystifying thing about it: for many of my patients, and for me as well, we become symptomatic when we eat gluten in the United States, but we can tolerate it in places such as Europe and Asia. Celiac is celiac, no matter where you are, but it appears that gluten-intolerance symptoms vary by geography. The classic pushback on this is, *It's just that you're relaxed when you're on vacation.* Indeed, the gut functions better when we're in a state of relaxation, but is that really a sufficient explanation for why the same pasta goes down easy on a honeymoon in Italy but leaves us bloated and groaning in our daily lives? A few years back, I put this theory to the test in my own body. I spent seven months living and working around the world. While in Italy, Greece, Israel, Hong Kong, Australia, and New Zealand, I ate *all the gluten*. And yet my skin was glowing and clear, and my digestion worked like clockwork. Then I touched down in Kauai, the northernmost island in the Hawaiian archipelago. Back to American soil and American agriculture. One bite of a pita, and all my gluten-intolerance symptoms came crashing back. I was still entirely in vacation mode—in the very *definition* of paradise, and yet I was doubled over in pain. What gives? My theory is that Roundup, the American-made pesticide we spray liberally on our wheat crop, plays a role in many Americans' gluten intolerance. This would make sense with the geography theory as Roundup's use is much more restricted in other parts of the world.[113,114] So, maybe all the American gluten intolerance is really just Roundup intolerance?

But what is it about Roundup? The active ingredient, glyphosate, has been demonstrated to cause intestinal permeability,[115] otherwise known as leaky gut. (The Environmental Protection Agency also suspects that glyphosate may injure or kill 93 percent of the plants and animals protected under the Endangered Species Act,[116] but that's a separate, devastating conversation.) Leaky gut allows some of the contents of the GI tract to enter the bloodstream, including portions of the partially digested gluten protein itself, provoking those with certain genetic susceptibilities to generate an army of antibodies targeted against gluten. Because our wheat crop is so heavily sprayed with Roundup, it's almost as if we consume gluten as a package deal—you get your gluten at the same time as a substance that causes leaky gut, which ensures that some gluten will leak into the bloodstream and generate antigluten antibodies. These antibodies can go on to attack the lining of the small intestine when the person eats more gluten, which then causes a vicious cycle of gut inflammation. (Some would argue that these antigluten antibodies can also attack thyroid tissue by way of something called "molecular mimicry," where thyroid tissue resembles the amino acid sequence of gluten just enough that the antigluten antibodies think they're attacking gluten, when in fact they're attacking our own thyroid glands.[117,118] This points to a possible relationship between gluten, thyroid conditions, and mental health issues such as anxiety and bipolar disorder, which at times are just thyroid dysfunction masquerading as mental illness.) Roundup, a probable carcinogen according to the World Health Organization,[119] is also theorized to harm bacterial diversity in the gut and thereby dysregulate the enteric nervous system.

You may be wondering, *OK, then can I just eat organic gluten that hasn't been sprayed with Roundup?* For some people, the answer is yes. It may be the case for you that eating a long-fermented organic sourdough, where the gluten is partially digested by the fermentation process and there is no trace of Roundup, will not present a problem. For others, after a lifetime of eating Roundup-laden bread, the gluten itself has become an inflammatory protein, even in its fermented and organic form, especially in the setting of ongoing leaky gut. That would imply not only that Roundup provokes our bodies to develop antibodies against gluten but also that these antibodies hang around for the long term.[120]

If you've discovered that your anxiety improves when you're off gluten and yet you're yearning to keep gluten in your life in some capacity, as long as you don't have celiac disease, experiment with organic sourdough and eating gluten when you're traveling internationally in places that don't spray their wheat crop with Roundup. As always, the ultimate test is *How do you feel?* If you can tolerate this without any physical symptoms or spikes in anxiety . . . Hallelujah! If, on the other hand, you find yourself feeling on edge, sad, anxious, or foggy or clutching your bloated abdomen within hours of eating gluten, be honest with yourself about your need to navigate this food with care.

Dairy, on the other hand, is a different story. We are all along a spectrum of dairy tolerance, and the different forms of dairy are themselves along a spectrum of potential for intolerance. At one end of the dairy spectrum, we have the forms most people tolerate well—products such as butter and ghee. A smaller segment of people do well with unpasteurized, full-fat, fermented sheep and goat dairy, such as kefir. On the other end of the spectrum, we have a glass of conventional, hormone-

and antibiotic-laden, skim cow's milk. As much as this has been marketed as a health food, it's highly processed and, according to animal studies, creates at least some evidence of an inflammatory reaction in the bloodstream.[121] Unless you already know you get a stomachache and loose stool every time you eat ice cream, do a trial to find out where you stand. Go off dairy for a month, monitor how you feel, and then add it back, again monitoring for physical and mental symptoms. Be honest with yourself about when your body says no. If dairy inflames you and you continue to consume it, it may be contributing to your anxiety.

This brings us to the final, troubling aspect of gluten and dairy: gluteomorphin and casomorphin. There's that root word again: *morphin*, like morphine. Indeed, gluteomorphin and casomorphin can behave like opiates in the human body.[122] When these components of wheat and dairy seep from a leaky gut into the bloodstream, they can cross the blood-brain barrier and potentially act on opiate receptors in the brain.[123,124,125,126,127] When someone with an inflamed gut eats pizza, it can be as though they just took a baby dose of morphine. This is one of the reasons we love pizza, but this can also make us feel fuzzy and groggy after a meal and can leave us antsy and craving more once the high has faded. So, that seemingly innocent muffin or cheese plate can leave us strung out, and this up-and-down swing can contribute to anxiety.

It's not always easy to know if you have a dietary intolerance. For some, it may be obvious, with symptoms such as bloating, gas, abdominal pain, diarrhea, constipation, or mucus in the stool after eating a particular food. For others, symptoms may occur downstream—such as acne, joint pain, rashes, eczema, congestion, post-nasal drip, itchiness, or migraines—or they may be even more subtle, such as brain fog or an uptick in anxiety. The best way to determine if you can tolerate gluten and dairy is to try a well-structured, monthlong elimination diet. Beware of hidden sources of gluten and dairy in foods such as soups, salad dressings, soy sauce, gravy, fried foods, oatmeal (unless it's marked gluten-free), and battered fish. Try to avoid slipping up during your elimination month, otherwise you'll end up with murky data. And when you finish, add the eliminated foods back into your diet systematically, one at a time. Observe how you feel generally, and notice, in particular, your anxiety levels during that month. If you go through withdrawal with worsening anxiety for the first few days of elimination and then your anxiety improves over the remainder of the month, these are clues that something you eliminated could be contributing to your anxiety. Once you can recognize how your body does with these foods, you have the power to make choices accordingly. What you do with that information is between you and your maker—sometimes ice cream is worth a little bloating, and sometimes a pancake isn't worth a panic attack.

I know that dietary restrictions get a bad rap. To onlookers, it might seem like a bunch of people being precious, obligating others to make a fuss over them, or like a sublimated eating disorder. In fact, dietary restrictions can be an act of courageous self-love and self-care—eating in a way that keeps you feeling well. In our society it takes commitment and a bit of swimming upstream. The key is to know your own body and make conscious, self-loving choices for yourself, day after day.

Herpes and Anxiety

When I work with my patients to investigate what may be causing their anxiety, I'm looking for patterns. One pattern I've noticed with a surprisingly large number of patients is that they experience a significant uptick in anxiety, panic, depression, and even hopelessness in the time right before and during a herpes outbreak. I suspect this is due to the corresponding uptick in inflammation as the immune system is activated by the flare-up.[128,129] As the outbreak and inflammation subside, their mood returns to baseline. If you have herpes and notice that your anxiety spikes when you have a flare-up, I suggest taking the steps outlined on page 120 to soothe the inflammation. Sometimes supplements or pharmaceuticals such as L-lysine, low-dose naltrexone (LDN), and even Valtrex can be necessary as well. If you notice your mood tracking with herpes outbreaks, start this conversation with your primary care doctor so you can get on your way toward quelling this underappreciated cause of brain inflammation and mental distress.

Chapter 9

Women's Hormonal Health and Anxiety

> Communities are only as strong as the
> health of their women.
>
> —*Michelle Obama*

A variation of false anxiety has existed for women in another realm throughout history—that is, not as a by-product of inflammation or lack of sleep, say, but rather because normal and justifiable behavior has been inappropriately labeled as mental illness simply based on gender. Around 1900 BC, it was suggested that women's moods depended on the position of their uterus. This thinking persisted throughout much of history, from the Greek physician Hippocrates (c. 460–c. 377 BC) first using the word *hysteria* to describe emotional volatility in women, also linked to the uterus, all the way on to Freud (1856–1939), the father of psycho-analysis, who asserted that a hysterical woman was a possessed woman, spurred on by a lack of "libidinal evolution."[1] And such sentiments are still echoing in doctors' offices today—distantly, perhaps, but resonating nonetheless. I have found that many physicians have a tendency to be dismissive or to make my female patients feel as if they're being high-maintenance or high-strung—modern-day euphemisms for "hysterical"—when they are either too inquisitive or vocal about their treatment plans.

I've been a witness to many such stories of bias. When a woman visits her primary care physician to discuss physiological issues, such as frequent stomach pains or neurological concerns, she may be told these symptoms are simply by-products of her anxiety. In other words, it's all in her head. For example, my patient Celeste, a fifty-nine-year-old marketing executive, reported tingling in her hands and sensations in her gallbladder for months to a variety of doctors; she was continuously told that these sensations were merely the products of anxiety and IBS. When Celeste finally found a doctor who took her complaints seriously enough to do an ultrasound and bloodwork, he discovered gallstones and a vitamin B_{12} deficiency. I would argue that the same sensitivity that makes people anxious also makes them more attuned to imbalances in their bodies. Or, in the words of Celeste, anxious patients are "reliable narrators," and we can generally regard their perceptions as useful information.

Another patient, Charisse, thirty-three, was made to feel that she was overstepping her bounds by attempting to take part in her medical care. Charisse was having irregular periods and made an appointment to see her gynecologist to discuss the issue. She spoke to me beforehand about not wanting to take any more medication—she had tapered off Zoloft a year prior—so we came up with a plan for her to advocate for herself in this regard. But at the appointment, when Charisse asked about addressing her irregular periods without medication, her doctor—

also a woman, by the way—asserted that birth control was the only reasonable choice and told her in a condescending tone, "You're welcome to go against my advice, that's your prerogative." Charisse, an African American woman keenly aware of the added layer of systemic racism in the doctor's office, soon capitulated but returned to my office feeling defeated. When she saw her gynecologist again, she explained that she was having frequent crying spells since starting the pill, and she was concerned that it was making her depressed. Instead of acknowledging the possibility that the pill may not be the best choice for Charisse, the doctor dismissed her concerns once again and suggested that she go back on Zoloft.

In addition to the silencing of women—even, as in the case of Charisse, by other women—there's another phenomenon I've observed, which is when female patients feel unable to communicate their needs, and their bodies begin to communicate on their behalf. That is, when a patient doesn't have a platform to express her feelings and needs (or she is systematically ignored), she may develop a tendency to "somaticize" deeply buried emotions into physical symptoms. I typically see this in female patients who are disenfranchised and oppressed in other corners of their lives—subject to societal disempowerment or conditioning, putting others before themselves in relationships, or otherwise subduing their needs in response to a culture that positively reinforces female martyrdom. Their bodies speak up when they can't, saying, *Something is not right here* or *I'm in pain.* Of course, this phenomenon is nothing new—the clinical exhaustion, conversion disorders, and pseudo-seizures of the days of yore are examples—but today I see it most often in the form of illnesses such as fibromyalgia, the basic symptoms of which are widespread pain and tenderness, and chronic fatigue syndrome. Both of these conditions have a valid material basis and cause very real pain and suffering. But they also exist at the intersection of physical and psychospiritual health. I believe that these ailments are a complicated commingling of the toxins of our modern world (dysbiosis plays a role in fibromyalgia[2]; mitochondrial dysfunction pertains to chronic fatigue[3]) *and* societal constraints—it is still, after all, less uncomfortable and stigmatizing to draw attention to physical pain than mental distress. In these instances, rather than physical symptoms being dismissed as "just your anxiety," the mental health struggles are manifesting as physical ailments. So, for these women—and they *are* mostly women, as fibromyalgia occurs nine times as often in women as in men[4]—their depression and anxiety often go unaddressed.

All of which is to say, if you are experiencing a health issue and you're met with a dismissive or invalidating attitude from your practitioners, do not let them shame you into deferential silence. Trust your body; *you* are the expert. Speak up, push back, self-advocate; rather than questioning yourself, question the system. We clearly still have further to go—and we all need to participate in continuing to bring our broken system forward.

Thyroid Dysfunction

An estimated twenty million Americans have some form of thyroid dysfunction,[5] most of whom are women. In fact, one in eight American women will develop a thyroid disorder in her lifetime;[6] currently, about 20 percent of women over sixty-five have a sluggish thyroid, and a large portion of those women are walking around undiagnosed.[7,8]

The thyroid is a small gland in your neck that powers the balance of energy throughout your body, and the thyroid hormones are primarily responsible for the regulation of metabolism. Hypothyroidism—a state of underactive thyroid activity—is an epidemic, and the medicine that treats this, called Synthroid, wins the prize for most prescribed medication year after year. Insufficient thyroid activity can cause fatigue, depression, constipation, hair loss, dry skin, brain fog, weight gain, difficulty losing weight, muscle aches, and exercise intolerance. If you are experiencing thyroid dysfunction, you might notice that the outer third of your eyebrow is very thin or missing. On the flip side, hyperthyroidism, a state of excess thyroid hormone, can generate high energy, insomnia, irritability, agitation, diarrhea, rapid heart rate, palpitations, feeling hot, excessive sweating, unintended weight loss—and anxiety.

A medical textbook would say that hyper- and hypothyroidism are two distinct, mutually exclusive disorders, but what I see in my practice is more complicated. Sometimes a state of excess thyroid activity precedes a state of underactivity. And most often people diagnosed with low thyroid have symptoms from both lists. I've certainly seen many patients diagnosed with hypothyroidism experience anxiety as a symptom of their thyroid illness.

If you suspect you may have thyroid dysfunction, then the first step is to book an appointment with a naturopath or functional medicine doctor to have your thyroid evaluated. (You read that right: most of my patients end up finding more relief from their thyroid symptoms when they work with a holistic practitioner than when they work with a conventional endocrinologist or primary care doc.) If, in the end, it turns out that you do have thyroid dysfunction, then you may want to take a combination of approaches to remedy the situation—ranging from thyroid medication to dietary changes and detoxification practices.

MENSTRUAL CYCLES AND MENTAL HEALTH

Although the menstrual cycle is typically equated with the shift in hormones that occurs just before a woman gets her period—and is usually associated with irritability—there are actually several other phases, and moods, that take place throughout. The follicular phase occurs in the first half of the month, after we start bleeding, when the predominant hormone is estrogen. During this time, as estrogen gradually ramps up, we might feel outgoing, confident, energetic, and social. Things don't get on our nerves quite so easily. Around ovulation, the higher levels of estrogen and androgen might contribute to an increased libido and higher energy levels. After ovulation, we're in our luteal phase, dominated by the hormone progesterone. At this point, as we inch closer to the phase when we bleed, we might feel annoyance and fatigue, perhaps preferring to stay home and take a bath rather than go out to a meet-and-greet.

We would do well to honor the body's request for rest at this time, as this is also when many women suffer from premenstrual syndrome, or PMS, where fluctuations in estrogen, progesterone, and serotonin can contribute to feelings of anxiety, depression, fatigue, food cravings, and sleep difficulties. PMS is a common, albeit uncomfortable, condition that affects an estimated three out of every four menstruating women in some form.[9]

However, many of my female patients experience an exaggerated and pathologic version of PMS because their hormones are off-kilter. In my practice, I see women who experience debilitating cramps, severe mood swings, and even suicidal thoughts in the days before their period. While some of this may be common, it is *not* normal. Instead, this is a sign of physical imbalance and a call to action. All too often, I observe a weary resignation to debilitating PMS among my female patients—as if it just comes with the territory of being a woman. But it doesn't have to be this way. Let's focus on how to get our hormones back into balance to eliminate this layer of amplified symptoms—or false PMS.

What I most commonly see in my practice is something called estrogen dominance. The existence of estrogen dominance is debated, but to me it seems biologically obvious in our modern world. The idea is that women's estrogen is too high and our progesterone is too low, which throws off the ratio of estrogen to progesterone. This causes a large hormonal crash in the luteal phase, which can look and feel like bad PMS or even premenstrual dysphoric disorder (PMDD), a more severe form of PMS that includes pronounced irritability, anxiety, and depression in the week or two before bleeding begins. Our estrogen is too high because we're constantly exposed to xeno-estrogens and endocrine disruptors (chemicals that mimic estrogen and alter the normal function of hormones, respectively) in the form of personal-care products,[10] makeup,[11] perfume,[12] cleaning products,[13] hand sanitizers,[14] plastics,[15] thermal receipts,[16] and pesticides.[17,18] And our progesterone is too low because of modern nutritional trends and chronic stress. To make progesterone, we need cholesterol[19] and a precursor molecule called pregnenolone. We come up short on cholesterol because it has been denigrated in the medical literature and the media, so we order egg-white omelets, mistakenly thinking we're doing right by our bodies, not to mention that some of us are even medicating our cholesterol away with statins. And we lack pregnenolone because it happens to be the precursor molecule to another important hormone in the body—cortisol, our stress hormone. So, every time we're stressed, our pregnenolone gets triaged toward making cortisol. You can imagine, with the chronic stress of our modern lives, how little is left over to make progesterone. This process is called "pregnenolone steal," and it leaves many of us with precariously low levels of progesterone.[20,21] Therein lies the perfect recipe for our imbalanced ratio of estrogen to progesterone, causing a steeper hormone crash in the days before our period, or, in other words: false PMS.

The good news is that we can take steps to bring our hormones back into balance, such as switching out personal-care products for natural alternatives, reducing exposure to pesticides and plastics, eating bitter greens to support liver detoxification of estrogen metabolites, and managing stress. The goal of balancing our hormones is not to deny the moods of the different phases of our cycle but to move through our cycle with more ease and less anxiety.

Additionally, I believe that when women bring their PMS into a more manageable realm of physical expression, they are also better able to reap the unexpected benefits of the luteal phase. Contrary to the prevailing cultural narrative that a woman's emotional insights during this time can be written off as irrational, my firm belief is that this time of the month allows access to deeply held convictions. It is not a time when we are less reasonable—it's a time when we have less tolerance for BS. So, we should honor the truth of the raw, irritable, and tender emotions of the luteal phase, and of PMS—it's a time for us to rest, go inward, and explore the revelations that may only fully surface during these crucial days of the month.

THE CONNECTION BETWEEN THE PILL AND ANXIETY

I often encourage my patients to stop taking hormonal birth control if I suspect that their mood struggles are linked to exogenous hormones—that is, hormones you take as opposed to hormones your body makes endogenously. The birth control pill can be an empowering choice—allowing women freedom and agency around their fertility and sexuality. That said, we know from recent research that exogenous hormones contribute to mood changes in the reproductive years, with the effect most pronounced in adolescents.[22] In my clinical experience, hormonal contraception can also play a role in anxiety. The pill may be the right choice for some, but I strongly believe that physicians must do a better job of fully disclosing the risks and side effects to patients, particularly as doctors are increasingly prescribing the pill for younger women. Recent research demonstrates a long-term association between adolescent oral contraceptive use and depression risk in adulthood, regardless of whether there is a continuation of oral contraceptive use. The authors of the study explicitly state that their findings point to the idea "that adolescence may be a sensitive period during which [oral contraceptive] use could increase women's risk for depression, years after first exposure."[23] In other words, starting hormonal birth control at a younger age seems to make mental health side effects more likely to develop, and they may last for years.

Not only do exogenous hormones pose an elevated risk of mood disorders but the pill also causes inflammation[24] and micronutrient depletion (particularly of the B vitamins, which pertain to anxiety and depression),[25,26] and it's hypothesized to induce changes to the microbiome.[27] It's also worth noting that over the long term, some oral contraceptives increase the risk of gallbladder issues[28] and certain autoimmune diseases.[29,30] One aspect of oral contraceptives that I've always found disturbing is that they increase the production of binding proteins, such as sex hormone binding globulin (SHBG), which then circulate in the bloodstream and bind up other hormones, such as androgens (what we typically think of as testosterone).[31] This has the effect of lowering available androgen, which can in turn impact energy, libido, and, yes, mood.[32]

I once treated a thirty-six-year-old woman named Naomi, who came to me on multiple medications for multiple diagnoses, including Wellbutrin for depression, Adderall for attention deficit, and Xanax for anxiety. She was also on the birth control pill. On our first visit, I asked her if she was open to getting off the pill. She said absolutely not, noting that the pill helped her with her acne and menstrual cramps, not to mention that she relied on it for contraception. We worked hard at addressing the root causes of her depression, anxiety, and ADHD. She changed her diet, started prioritizing sleep, built small amounts of exercise into her daily schedule, and even established new boundaries at work. Ultimately, we got her off the Wellbutrin and Adderall, which she found very freeing. She celebrated no longer having an Adderall "comedown" every afternoon, where she would find herself exhausted, ravenously hungry, and irritable. She also made improvements in her depression and ability to focus. But she was still struggling with background anxiety. Then, a few years into our treatment, she and her boyfriend broke up, and she decided to quit taking the pill.

The effect was immediate and marked. Naomi's anxiety

simply went away. Her panic attacks stopped, and she just felt *good*. Life still had stress, but she found herself coping with the stressors coming at her with a resilience neither of us had seen before. She continued to thrive—until she got into another relationship and decided to get an intrauterine device (IUD) that releases progesterone. It's suggested that this progesterone has only a "local effect," meaning it only impacts the uterine lining. But in reality, if progesterone is getting released into your uterus, it's getting into your bloodstream and going everywhere, including your brain.

Though I had tried to steer Naomi toward nonhormonal birth control options—like the copper IUD, the fertility awareness method, and condoms—she had pushed back: her friends told her "horror stories" about heavier periods with the copper IUD, she didn't trust herself to do fertility awareness reliably, and condoms were a nonstarter with her boyfriend. Within a week or two of getting the progesterone-eluting IUD, Naomi was back to being the same ball of anxiety that she'd been when I met her.

I pointed out the chronological connection between the IUD placement and her resurgence of anxiety. Naomi responded with a frazzled recounting of how work was crazy these days. Sure, work was crazy. Work was *always* crazy. But the same crazy job had gone from being intolerably stressful to tolerably stressful and now suddenly back to intolerably stressful again. The job hadn't changed; Naomi's hormones had. After a few months, Naomi's anxiety reached such a fever pitch that she was willing to try anything, including having her IUD removed. Within weeks of having it removed, her anxiety resolved again.

It was clear that hormonal birth control was at the root of Naomi's anxiety. I took a step back to reflect on her medical history. When I revisited my notes from our initial session, I noticed something I had overlooked: Naomi went on birth control for acne when she was sixteen, at around the same time she had first been diagnosed with depression and anxiety. This was a moment of clarity for me: *all* of Naomi's mental health diagnoses had begun just after she started birth control as a teen. Sometimes new symptoms are actually the side effects of new medications. Nobody—not even me right away—connected the chronological dots between the onset of Naomi's depression and anxiety with the initiation of hormonal birth control.

I am overwhelmed with sadness and anger when I think about the ways this has impacted Naomi's life for the past twenty years—the suffering it has caused, the impact it has had on her relationships and career and on her sense of herself. And Naomi is just one of many female patients whose anxiety has tracked with hormonal birth control. Her anxiety disorder dates back to the first time she was prescribed the pill—and tracks perfectly with whether she is on or off it. And yet all she has ever been told is that she has mental illness. Thankfully, Naomi plans on using nonhormonal contraception from this point onward. If this story reminds you of your own mental health journey, consider this: before defining yourself by a diagnosis, first rule out whether you are suffering from medication side effects that *mimic* mental illness—hormonal birth control being a common culprit.[33] These labels, diagnoses, and seeming destinies that can become so tied up with our identities are sometimes just temporary imbalances, subject to change.

On PCOS and
the Idea of a "Cure"

When I posted on social media about my functional medicine–based approach to resolving polycystic ovary syndrome (PCOS), a hormonal disorder causing infrequent or prolonged menstrual periods and symptoms of excess androgen, I received some criticism suggesting that I was misleading people because there's no "cure" for this.

The broad idea of a "cure" is an outdated notion that implies that scientists will have a eureka moment and come up with a solution that *solves* a disease. But disease processes are often far more complex than that, and prevention will always be preferable to treatment. Our health is determined by a combination of genetic predispositions and environmental influences (e.g., diet, lifestyle, stress, and exposures). And while we don't control our genes, we *can* control environmental influences and gene *expression*.

The genetic vulnerability to PCOS is as old as time,[34] but the prevalence of PCOS is increasing,[35] and this is accounted for by environmental influences.[36,37,38] The root causes of PCOS include high levels of cortisol,[39] insulin resistance and obesity,[40] and chronic inflammation.[41] When we ditch the default modern diet and lifestyle that are making so many of us sick, the signs and symptoms of PCOS often abate. We can debate the semantics—cure, reverse, heal, resolve—but if you go from irregular periods, infertility, and hirsutism to a state of hormonal balance, periods like clockwork, and a restoration of fertility, you probably don't care what word we use.

FERTILITY AND ANXIETY

We live in strange times. In our day-to-day lives, we are immersed in a sea of factors that affect female and male fertility—from chronic stress, pesticides, and endocrine disruptors to cell phones in pockets affecting sperm count.[42] And yet our culture remains myopically focused on one issue when it comes to fertility: a woman's age. There's an acute and undeniable emphasis on a woman's "biological clock," and although we're now more able to control our fertility than ever before, we're also now more at risk for anxiety about it. Workplaces demand long hours, and working moms juggle being present for their families (and being the first number called when a kid gets a fever at school) with the risk of getting "mommy tracked" in the prime years of generativity. So, many of us use hormonal birth control to avoid having children when we're not ready and then end up needing fertility treatments to conceive children when we are. Some of us freeze our eggs in an attempt to escape the modern fertility bind. These can all be empowering choices, but we should bear in mind the tightrope we're walking as a result: there's the pressure to procreate, the pressure to not let our childbearing get in the way of our work, and the pressure to not wait so long to have kids that we have higher-risk pregnancies or miss our chance to have a family altogether. And even if you have zero interest in having kids, that comes with its own pressure to answer incessant, unsolicited questions about this choice. From the strain this puts women under to the dizzying array of time-sensitive decisions that need to be made along the way, my female patients of childbearing age are increasingly anxious.

And it's not as though the anxiety ends when you do get pregnant. Instead, it is as if the brain seems to grow an additional fold dedicated to lying awake at night in nervous anticipation of any potential bad outcomes (which only continues, with gusto, in motherhood). In fact, a 2018 study, observing 634 pregnant women during their first trimesters, found that the prevalence of "high state anxiety" was 29.5 percent.[43]

There are also, of course, particular challenges around pregnancy loss. While great strides have been made in recent years to destigmatize miscarriage, it remains a part of a woman's health journey that lacks adequate resources, understanding, and support. Partly this stems from the fact that for so long, women have felt discouraged from sharing these experiences. And yet miscarriage occurs in 8 percent to 15 percent of clinically recognized pregnancies (and 30 percent of all pregnancies),[44] with the great majority of the risk falling in the first twelve weeks. As such, many women are uncomfortable sharing news of their pregnancy until they've safely made it to the second trimester. I understand the desire for privacy and boundaries, though I wish we felt safe enough to share this news earlier so that fewer women would suffer in silence. I feel so strongly about this issue, in fact, that I took the risk of sharing my own pregnancy news early. At ten weeks into my second pregnancy, I posted the news on Instagram.

And then, at eleven and a half weeks, I miscarried.

As a physician, I well understood that miscarriage is common, normal, and natural. A significant percentage of embryos have genetic errors (or typos), and miscarriage is the system of checks and balances that prevents nonviable embryos from making it past the early stages of pregnancy. This system has been in place throughout human evolution. While genetic errors increase with age—maternal and paternal age—they can occur at *any* age. Also, chromosomal abnormalities are not the only cause of miscarriage; placental placement and viability, blood clotting tendencies, and other more intangible factors are also at play. Miscarriage can happen to anyone, I knew, and it doesn't mean that there's anything wrong with you or that you did anything wrong—and yet all that knowledge didn't make the experience any easier.

I spent a lot of time sitting in the raw emotional discomfort. And as the dust settled, I felt a combination of grief and peace. I still hold these two seemingly contradictory feelings at once. Miscarriage is a steep loss, and I've treated many female patients grappling with anxiety and depression after losing a pregnancy. Indeed, I have worked with some women who have experienced several miscarriages in a row—after having undergone seemingly endless rounds of IVF—and find that the ongoing grief and precipitous hormonal crashes morph into more serious forms of anxiety and depression. It is perfectly normal to experience a disorienting mixture of devastation, hopelessness, anger, numbness, and even relief all at once. There is no right or wrong way to feel.

My experience with pregnancy loss revealed to me that the medical field has a poor understanding of what a body and mind need after miscarriage. I was bleeding heavily, light-headed, and depleted. My uterus was beginning to prolapse (i.e., fall out of position). My body was asking for rest. Instead, I was advised to run around and get additional sonograms, blood tests, and IV fluids the morning after my miscarriage. I didn't need to wait in triage only to perch uncomfortably on an ER gurney; I needed to sip tea while lying on a couch under a blanket, sifting through what had just happened. After a miscarriage, it is of course important to heed medical advice, as there can sometimes be serious complications, but it's also necessary to listen to our bodies and advocate for our mental and physical well-being.

I was lucky enough to have Kimberly Ann Johnson—doula, trauma expert, and author of *Call of the Wild*—on speed dial.

When I sought her support during this experience, she explained that "miscarriage is essentially a postpartum period," proportional to the amount of time you were pregnant. An important difference between a postmiscarriage and a postpartum period is that, at least in our culture, we don't have a tradition of rest and time off from work after miscarriage. How could we when first-trimester pregnancies and miscarriages largely go undiscussed? (Given that our institutions barely acknowledge the necessity for time off and recovery after childbirth, it would seem we're a long way from acknowledging the time needed for recovery after miscarriage.)

Once I had spent a few weeks healing and reflecting, I returned to Instagram to share the news of my miscarriage. The post was met with an incredible amount of warmth and support, which of course was much appreciated, but most importantly, it seemed to strike a resounding chord among those who had been through a similar experience. Women came forward to tell me about their varied experiences with pregnancy loss, and I recognized what an honor it was to hold space for so much tender processing. I heard a collective sigh of relief from women in all corners of the world—essentially saying, *I'm so glad we can talk about this.* As a society, I think we're ready to carry the heaviness of pregnancy loss. My hope is that we can all begin to discuss our pregnancies whenever we feel ready, normalize the conversation about miscarriage, release self-blame, and systematize the care and support that we should receive during this painful chapter of our lives.

POSTPARTUM ANXIETY

Of all developed countries, the United States offers some of the worst conditions for mothers. Devastatingly, in 2018, there were approximately 660 maternal deaths in the United States (a rate of 17.4 deaths per 100,000 pregnancies)—ranking us last overall among industrialized countries.[45] Furthermore, according to the Centers for Disease Control and Prevention, Black women are three times as likely as white women to die from a pregnancy-related cause.[46]

And once we actually make it across the threshold of child-birth, we begin the journey, along with all women the world over, with unequal pay at work and, on average, an unequal division of unpaid household labor.[47] Many of us—around 11 percent of the U.S. population[48]—are without health insurance to cover prenatal care and childbirth. Mothers who work outside the home are then faced with inadequate or nonexistent parental-leave policies, unaffordable childcare options, and potential job insecurity. It's no wonder that an increasing number of new mothers are anxious. In fact, over the years from 2017 to 2019, of the new or expecting mothers who took a Mental Health America screening test, 74 percent scored positive for moderate to severe for a mental health condition.[49] The prevalence for anxiety disorders in the postpartum period is 17 percent, exceeding the rate of postpartum depression (4.8 percent).[50] And yet awareness of postpartum anxiety (PPA) has greatly lagged behind that of postpartum depression (PPD). In part, this is because PPA is a more recent diagnosis as compared to PPD,[51] but also the public remains less informed about this form of anxiety, tending to assume depression is the mental health issue women generally struggle with postpartum. In fact, PPA can occur on its own or alongside PPD, causing women to feel fearful and overwhelmed. Some of the other symptoms of PPA include racing thoughts, an inability to sit still, and physical symptoms such as dizziness, hot flashes, or nausea. It is worth noting that the sleep deprivation that occurs with newborn care can also contribute to—or compound—these feelings.

As if these cultural, social, and financial factors don't pose a great enough challenge, the physical realities of growing a baby in the body, going through labor, and recovering from labor, plus or minus nursing, all while sleep deprived and strapped for time, are a perfect recipe for nutritional depletion. I believe this depletion of the body's nutritional stores is a significant and underappreciated root cause of PPA, meaning that the path to recovery may be as much about replenishing nutrients as it is about therapy and processing the transition into motherhood.

There is another window of health vulnerability, albeit of a different sort, in the postpartum period. *During* pregnancy, our bodies naturally dial down immune activity so as not to attack the baby and placenta, which could be perceived by our immune system as "foreign."[52] Many of my patients with autoimmune disease experience relief from their symptoms during pregnancy. In the postpartum period, however, the immune system comes back online precipitously, making new moms particularly vulnerable to inflammation and the development of autoimmune disease.[53]

Consequently, it is imperative to be gentle with your immune system in the postpartum period, lest you tempt it into a state of dysregulation. If you know you have a dietary intolerance, do your best to avoid it in those early weeks after giving birth. Down the road, you can make a judgment call about whether that croissant is worth it. But the postpartum period is a sensitive few months when you're particularly susceptible to developing or exacerbating autoimmune disease. And, of course, excessive inflammation directly contributes to anxiety. It follows then that by minimizing inflammation, you help minimize postpartum anxiety.

New moms need a village of support, generations of wisdom, lactation guidance, warm broths, nutrient-dense foods, and someone to listen as they process their birth experience and the changes to their bodies, lives, and identities—not to mention someone to prepare those warm broths and hold the baby while they take a shower and a nap. In our current system, what they get instead, on top of a major life transition and a physical ordeal, are the perfect conditions for anxiety. Let's turn up the volume on this conversation and recognize the immense need and the actionable steps we can take to support ourselves and other mothers.

Chapter 10

The Silent Epidemic

Never again will a single story be told as
though it were the only one.

—*John Berger*

Although I believe that physicians take the Hippocratic oath to
"do no harm" seriously, I also believe that today we have created
an unintentional crisis in care in which the medications that
doctors so often prescribe create their own type of false anxiety.
Nowhere is this truer than with mental health. And this is where
psychiatry finds itself today: labeling people with diagnoses as
if it were a genetic destiny, while offering medications that can
overlook the root cause of mental illness and, in some cases, exacerbate the very problems they are intended to fix. Meanwhile,
we have also neglected to address the side effects and long-term
consequences of these medications as well as the sometimes excruciating withdrawal process that may occur if and when the
patient decides to stop taking them.

These circumstances are especially meaningful now given

that we're at an all-time medicated high: the United States is the most medicated country in the world, with one in two people on a prescription medication, the most prevalent of which is antidepressants for Americans under the age of sixty.[1] Not to mention that the number of people taking psychiatric medications jumped notably during the COVID-19 pandemic: in the United States, prescriptions for antianxiety medications—such as the benzodiazepine class of drugs (or benzos), including Klonopin, Xanax, and Ativan—rose 34.1 percent from mid-February to mid-March 2020, while antidepressant prescriptions rose 18.6 percent in that same time period.[2]

The false anxiety caused by these psychiatric medications can occur through several different pathways: certain meds, such as the stimulants Adderall and Vyvanse as well as the antidepressant Wellbutrin, directly exacerbate anxiety through the modulation of dopamine and stress hormones, such as norepinephrine, and create an activated state that can feel like anxiety; other medications, such as SSRIs and benzos, can leave people in "interdose withdrawal," meaning a state of relative withdrawal as the body hits its pharmacological nadir, or low point, in between doses, which creates a chemical fallout where the body is anxiously awaiting its next dose. Perhaps the most substantial consequence, however, is that benzos can impact GABA-receptor expression in the long term, making it difficult for a person to feel calm on their own without taking more.

On that last point: the steep rise in people taking benzos, particularly, has created a troublesome predicament that is largely overlooked by the medical world. Keith Humphreys, a psychologist and professor at Stanford University, even referred to benzos as "the 'Rodney Dangerfield' of drugs" because they don't get the attention they deserve, given how addictive and destructive they can be. "Maybe people think that since they come from a doctor they can't be all that bad," he reasoned.[3]

One of the biggest problems with benzos is their impact on GABA-receptor behavior. As you'll recall, GABA is the primary inhibitory neurotransmitter of the central nervous system. It's the one that says, *Shhhh, there, there, it's OK*, creating our capacity to feel calm and at ease.[4,5] Unlike the scattershot target of SSRIs—which are meant to aim at one neurotransmitter as a bull's-eye but disrupt many in the process benzos offer a straight shot. They act directly on the $GABA_A$ receptor, creating the effect of a rush of GABA in our synapses—which, of course, can feel amazing.[6] For many, upon swallowing a benzo, the world feels suddenly like a peaceful and nurturing place. I completely understand why many people are reluctant to reconsider their relationship to these medications. When you're struggling with anxiety, taking a benzo feels like a warm hug. Who doesn't want a quick and reliable way to feel comforted in these increasingly stressful times?

Unfortunately, however, that is not the end of the story. Benzos can really only work their magic in the short term, and they can leave you worse off than when you started. The reality is that the body is hardwired for survival—not for feeling calm. So, when we create a rush of GABA in our synapses, the body responds by trying to reestablish homeostasis—or the original state of balance. It says: *That's too much GABA. What if a real threat did come along? We'll be too chilled out to care—and we won't survive.* So, it responds to benzos by downregulating the GABA receptors.[7] After this, it's as if our brains can't *feel* the GABA anymore. And when the medication wears off, we are left with normal amounts of GABA, but abnormally few receptors. This causes a state of relative GABA withdrawal, and over time, benzos can have a cumulative impact on GABA signaling, creating a significant withdrawal state—a feeling that can range from anxious and irritable to excruciating or, according to one of my patients, like being dragged through hell by the hair. Benzo tolerance—when the body acclimates to benzos and requires more to get the same effect[8]—leaves many feeling more anxious than before they began taking the medication in the first place. Indeed, benzos have been shown to exacerbate anxiety—the exact issue they're being used to treat—in the long term.[9] You can think of benzos as a Band-Aid that leaves behind a bigger cut than the one it was being used to cover up. And that new cut is essentially benzo-generated false anxiety.

But it's well known that benzos are also habit-forming—they create "physical dependence," a medical euphemism for addiction. So, often, when patients return to their psychiatrists after two weeks for a refill (because now this pill is the only thing that will ease their anxiety), the doctors will sometimes shift to a stance of withholding and shaming. Now it's, *You have to get off this stuff—it's not good to be on it long term.* "You might expect doctors would take the difficulty of coming off these drugs into account," as the pseudonymous psychiatrist and blogger Scott Alexander points out, "but you might expect a lot of things from doctors that don't always happen."[10]

WITHDRAWAL: THE SILENT EPIDEMIC

I have witnessed firsthand how excruciating it can be for some people to get off psychiatric medications. In fact, I've treated so many people struggling with withdrawal from psych meds that I view it as a silent epidemic. Indeed, a 2019 analysis by

Drs. James Davies and John Read, both researchers based in London, found that 56 percent of people who attempt to stop taking antidepressants experience withdrawal effects and, of that group, 46 percent describe the symptoms as severe.[11] Which means that the recent meteoric rise in numbers of people taking both antidepressants and benzos that occurred during the pandemic will likely have a drastic impact when some of those millions decide that they'd like to come off them.

If meds were easier to discontinue, I would be more inclined to prescribe them. But there are legions of people out there in the midst of withdrawal—experiencing such symptoms as insomnia, irritability, depression, anxiety, brain fog, fatigue, nausea, panic attacks, and brain zaps (which feel like an electric shock in the brain)—who feel alone in this experience and struggle to find support or appropriate guidance from doctors. What is most concerning to me is that many people experience suicidal thoughts in the midst of withdrawal. I have even seen suicidal ideation appear for patients who had never experienced such thoughts prior to withdrawal.

Benzos pose a particularly troubling withdrawal predicament. Since GABA is the essence of feeling OK, GABA withdrawal is the quintessential feeling that *nothing* is OK. I've been at this for some time, and of all the withdrawal states I've witnessed, from Adderall to heroin, benzo withdrawal is arguably the most painful for my patients. I have seen them irritable, anxious, unable to sleep, in a panic spiral, despairing, suicidal, and feeling as if they want to crawl out of their own skin.

Indeed, after three or four months of daily use, benzos can be immensely challenging for some people to quit. It's difficult to predict who will be able to walk away unscathed and who will get hooked. No doctor intends to addict their patients to benzos, but it happens so often because these meds create the need for themselves. Once you're in withdrawal, for instance, you need to take more benzos just to feel normal again—that is, back to the old anxious way you felt before you even started the medication—let alone relaxed. And, unfortunately, the longer and more consistently people take these meds, the harder it is for their GABA receptors to start working normally again. How can this be the best way to help people with anxiety?

In my own approach to patients with benzo withdrawal, I focus on helping to build back up natural GABA activity (as we discussed in chapter 7) by improving nutrition and sleep; decreasing alcohol consumption; and instituting breath work, meditation, yoga, chanting, or acupuncture—all of which prompt natural GABA resilience. While these steps are helpful, I will also acknowledge that it's still not always easy.

This situation is made even more difficult by the fact that psychiatrists are not even taught *how* to help people get off these drugs; in fact, most deny the very idea that getting off meds such as SSRIs can cause withdrawal. In our training, we're instructed that if a patient becomes symptomatic soon after getting off their antidepressant—exhibiting signs of increased anxiety, panic, insomnia, crying spells, or worsening mood— this should be considered a *relapse*, as opposed to *withdrawal*. While of course a relapse is possible, I find it difficult to verify whether somebody is in a relapse during the first few weeks after medication discontinuation, when their brain chemistry and receptor expression is in the process of re-equilibrating. If a person with a history of depression woke up the morning after a cocaine bender and felt dysphoric and lethargic, would we call that a relapse of their depression, or would we recognize that their mood was temporarily affected by drug withdrawal? Make no mistake, antidepressants are powerful psychoactive substances, and their discontinuation leads to a real withdrawal. We currently have an epidemic of people experiencing withdrawal symptoms upon discontinuation, without knowing they should attribute these symptoms to their medication. Instead, they either blame themselves or their life circumstances or simply see it as a relapse—and an indication that the medication was helping.

For instance, I once had a patient named Tova who had been moderately depressed and anxious for years—from her adolescence into her twenties—until her primary care doctor put her on Lexapro when she was twenty-five. Though she didn't experience an immediate improvement in her mood, she did notice, over time, that she was crying less and seemed better able to function in her work and personal life. Tova soon felt so well that she began to question whether she still needed Lexapro, so she stopped taking it cold turkey, and then, as she put it, "all hell broke loose." She felt irritable and anxious, she snapped at her roommates and her mom, and she couldn't sleep to save her life. So, she went back on her medication. These discontinuation cycles—in which Tova would begin to feel pretty good, stop taking her medication, and then grow increasingly unhappy—occurred repeatedly over the course of several years. And each time, Lexapro would descend as if it were a gift from the heavens, restoring Tova to normalcy and calm. By the time Tova came to see me, she was fully crediting Lexapro with saving her life.

Tova's clinical history, however, suggested to me that the root cause of her mental health struggles might be something other than a straightforward chemical imbalance. Furthermore, I suspected that what was really happening each time she went off and on her medication was that it was rescuing her from withdrawal. So, while Tova was praising Lexapro for saving her from depression and anxiety, I believed it had initially blunted her emotions enough to seem effective, and after that it was simply relieving its own withdrawal. Tova noted that the Lexapro decreased her libido, yet when I began a conversation about one day tapering off Lexapro, Tova was understandably defensive. She felt judged, and she braced against the idea of my denying her this most essential support. I assured her that I understood the immensity of her struggle and the value of something that brings her relief. And, of course, there is zero stigma or shame when it comes to managing mental health with medication in my practice—but there is always an acknowledgment of the trade-offs that come with meds.

Over the course of several months, Tova and I explored what the roots of her depression and anxiety might be. We began to see patterns that suggested her mood struggles could be based on a combination of dieting and calorie restriction, insufficient healthy fats in her diet, caffeine sensitivity, side effects of the birth control pill, compromising her beliefs to please others in her life, and chronic Lyme disease. One by one we worked on each of these issues, correcting them at the foundation. With these changes, Tova felt well enough to attempt a gradual taper off her Lexapro, reducing it by 10 percent per month, rather than the cold turkey method she had been doing on her own. We also supported her taper by incorporating better nutrition, plentiful rest, meditation, and the use of an infrared sauna to support detoxification as her Lyme infection was being addressed. We agreed on ways for her to set boundaries in relationships and advocate for her own needs. After about a year, Tova was free of Lexapro, as well as depression and anxiety.

WITHDRAWAL PRIMER

When I'm supporting patients who are attempting to taper off psychiatric medications, I treat each case as an entirely unique situation, as I find medication discontinuation to have a lot of individual variation. However, there are some reliable approaches I employ for all patients while they taper: setting the initial rate of decrease of the medication to 10 percent, supporting detoxification, supporting the nervous system, and creating space for emotional release.

I typically recommend decreasing the medication dose at a rate of 10 percent of the most recent dose per month to minimize withdrawal symptoms.[12] To make this possible without my patients needing to cut pills at home (which can be an inexact science), I work with compounding pharmacies—these are pharmacies that can use generic medication powder and weigh out exact amounts to make customized pills—instead of being beholden to the commercially available pills from the manufacturer.

When patients are apprehensive about tapering because they have been traumatized by a previous withdrawal, I point out that tapering slowly is a considerably less painful process than what they might have experienced when discontinuing cold turkey. Meanwhile, there are still other patients who are eager to be off medication, and they sometimes feel that decreasing by 10 percent per month will take too long. (The process typically takes about a year, depending on whether we pause or adjust the rate.) For them, I emphasize that it's critical to go slowly but successfully rather than quickly but unsustainably, which often leads to patients going back on their meds to deal with their withdrawal symptoms. I also inform my patients that as they get down to lower doses, it's not really that they're still "on" their medication, because they are no longer on a therapeutic dose—they are simply on *enough* medication that their body won't be thrown into withdrawal. Indeed, in the last several months of a taper, patients are effectively "off" medication and just staving off withdrawal with small amounts of the med.

During the taper process, it's important to support the body's mechanisms for detoxification in order to help clear the breakdown products of medication metabolism. I usually recommend staying hydrated, taking Epsom salt baths, and, if it is logistically and financially feasible, using an infrared sauna regularly. A subset of my patients has found coffee enemas to be particularly helpful. These involve using room-temperature coffee inserted in the rectum to elicit a thorough evacuation of the large intestine and promote detoxification. While this is nobody's favorite activity, I have witnessed it rescue many of my patients from challenging tapers.

I have consistently found that the nervous system gets fairly disoriented during a taper, which can show up as mood swings, irritability, insomnia, and feeling easily overwhelmed, so I also recommend that patients practice daily meditation and breath work in order to nudge themselves back to a state of relaxation at least once per day. And, as always, I encourage patients to prioritize good sleep habits and nourishing food—both critically important to allowing the body to rebuild receptors and neurotransmitter stores in the brain.

Finally, most people experience a significant degree of emotional release during a taper. This is expressed in a range of ways by my patients—some feel waves of sadness or anger; others cycle in and out of states of despair. On a practical level, this is probably due to their neurochemistry re-equilibrating. On a psychospiritual level, however, I believe this is the recapitulation and release of emotions that were blunted by the medication. If someone lost a loved one while on meds, for example, I often witness a delayed grieving process during the taper. I approach the emotional release of medication discontinuation like a midwife approaches a healthy labor: I recognize that it is painful, but I don't react with fear. I sit steadily by, holding space for everything that comes up, while repeatedly reassuring my patients, *I'm right here with you, and you've got this*. It's a process, but it's also reliable in that I have consistently seen my patients ultimately find their balance, feeling steady on their feet once they've adjusted to their new emotional lives.

INFORMED CONSENT

At the end of the day, the decision of whether or not to choose medication should be reached after a thorough, thoughtful conversation—or series of conversations—between patient and prescriber. This is true informed consent. Unfortunately, these types of exchanges are increasingly rare in today's world of rushed appointments. In many cases, patients are not fully informed about the potential side effects of psych meds—commonly including weight gain, digestive disturbances, and low libido. What I find most troubling, however, is that there is almost never a discussion of the withdrawal that can occur if and when one wants to stop taking the medication. I believe that all doctors have an obligation to help patients weigh this consideration with the benefits of medication before prescribing that first pill.

Last, it is also critical to point out: If you're feeling discouraged by this discussion—feeling stuck on meds or anticipating a difficult withdrawal—know that it's never hopeless. Our brains are, indeed, quite plastic and adaptable. This is what brains do: *they learn*. They learn to rely on meds, and they can also learn to recover and rebuild after meds. There is always hope.

Chapter 11

Discharging Stress and Cultivating Relaxation

These mountains that you are carrying,
you were only supposed to climb.

—*Najwa Zebian*

The false anxiety generated by your body's stress response can be prevented in part by making changes to your diet and daily habits. But there are a couple of other important considerations to factor into your body's physiological experience of anxiety: cultivating the relaxation response and completing the stress cycle.

The first—cultivating the relaxation response—is a bit like taking a multivitamin that fends off anxiety. This practice helps your body raise its stress threshold, making it harder for anxiety to send you into the stress cycle in the first place. The second—*completing* the stress cycle—acknowledges that there are unavoidable stresses in life and that it's important to expel our charged energy so that our bodies can return to baseline. Ultimately, the more we can do on a daily basis to dial up the parasympathetic nervous system responsible for the relaxation response and dial down the sympathetic nervous system responsible for the stress response, the less anxious we'll be.

THE RELAXATION RESPONSE

We've all been shouted at to "just relax!" at some point, and so we all know this command is useless, often only amplifying whatever state of anxiety or dysregulation we're already in. But there actually *are* ways to use science to enhance the body's capacity to return to a state of calm. Carving time out for these practices daily is a great way to keep overall anxiety levels in check.

Our autonomic nervous system is where all the action lies. This part of our nervous system has two branches: sympathetic and parasympathetic. You can think of these as our stress response (sympathetic) and our relaxation response (parasympathetic). The sympathetic state is what we've been discussing so far—it directs our stress response, determining when we fight or flee. Communicating through the stress hormones cortisol, adrenaline, and norepinephrine, the sympathetic system tells us when we're not OK and tries to get us to do something about it, with an eye toward prioritizing our survival in the moment. For many of us, this state feels synonymous with anxiety. The relaxation response, however, is the converse of the stress response. It occurs when the parasympathetic nervous system "purrs to life" with neurotransmitters such as acetylcholine, serotonin, and GABA, as Herbert Benson, MD, director emeritus of the

Harvard-affiliated Benson-Henry Institute for Mind Body Medicine, puts it in his book *The Relaxation Response*, prompting the body to take time to rest, digest, and repair.

The stress and relaxation processes are mutually exclusive the overriding tone of your nervous system can't be in both states at the same time. This means that as long as you can tip your body into a relaxation response, you are not only avoiding a stress response but also raising the body's threshold for stress. It's as if your nervous system has a zero line, and the more you lift it above that zero line into relaxation territory, the farther you have to travel to dip below zero into a stress response. So, a few minutes of cultivated relaxation can make it less likely that you'll drop into a state of anxiety on any given day.

How do we spend more time in a relaxation response? Well, to start, we can get there the old-fashioned way: that is, by literally being relaxed (imagine that!). This generally requires that you get enough sleep, your body is replete with the nutrients it needs, all is quiet in your gut, there's no unresolved trauma, the world is safe for you and your family, you're not grieving, you have enough, you know you are enough, and you're not sheltering in place during a pandemic. All of which is to say: being relaxed does not come easy these days. So, we are going to need a few hacks.

First, it's important to understand that the mind-body connection is a two-way street. When our mind is relaxed, it tells our nervous system to slow and deepen our breathing. Our jaw relaxes, our digestion revs up, and the blood vessels in our hands and feet dilate. The end result is a feeling of bodily calm. Meanwhile, a relaxed body also sends a signal to the brain to follow suit, turning our thoughts toward ease, gratitude, and wonderment. You'll recall that a large portion of communication along the vagus nerve is *afferent*—or sensory data—meaning that it's communicating *from* the body *up* to the brain.[1,2] As such, we can trick our gullible brains into relaxation by creating some of the conditions of a relaxed body.

There are a number of ways to generate the physical conditions of relaxation in the body, including yoga, meditation, Tai Chi, acupuncture, craniosacral therapy, reiki, yoga nidra, breathing exercises (see below), and progressive muscle relaxation, just to name a few. It's also possible to directly stimulate the vagus nerve with practices like gargling, chanting, humming, breath work, cold showers, and even cold-water plunging.

Short-Circuit the Stress Response

A particularly effective entry point for cultivating the relaxation response is altering the way we breathe. When we slow our breathing, our diaphragm will send a message up to our brain: *By Jove! I can't believe I'm saying this, but for once it seems that we're relaxed!* In other words, breathe like a relaxed person, and your body will tell the brain you are a relaxed person.

Breathing exercises that lengthen the exhale relative to the inhale can induce a relaxation response in the body because taking longer exhales mimics what your body does when it is genuinely relaxed.[3,4] You can try this right now. Take one minute to lie on your back, placing your hands on your belly, and breathe with the following pattern: inhale to the count of 4, hold for 7, and exhale to the count of 8. Let it feel calm and easy. A count is not necessarily a second—it should be whatever unit of time allows you to do the exercise without straining. After you do a few rounds, check in with yourself. How do you feel? Did you experience any shift in your anxiety level?

Note: sometimes when we begin to pay attention to our breath, we can unwittingly hold it in funky ways and thus strain our breathing and trigger anxiety. If your breath becomes strained, simply focus on *allowing* it to be easy and natural, rather than *forcing* it to follow a certain pattern. If you need to reset your breath, you can always take what's called a cleansing breath—inhale through the nose and exhale through the mouth with an audible sigh.

The breath also provides an entry point of a different sort into managing false anxiety. Most often, I use breathing exercises to help patients break the habit of shallow, rapid breathing and cultivate a more relaxed breathing pattern. But some of my patients are unable to breathe properly because of structural or physiological issues, and this can be a profound contributor to anxiety. Remember that anxiety is frequently the consequence of the body being in a stress response—and what could be a stronger signal for stress than a body thinking it's ever so mildly suffocating?

If you suspect you have barriers to proper breathing—perhaps you breathe through your mouth or snore heavily at night—it's worth getting to the bottom of this, whether that means seeing a functional manual therapist to widen your hard palate and help open your nasal airways, or an osteopath to help with proper diaphragm function, or being evaluated for sleep apnea. It may be that you simply need dust mite covers on your pillows. Whatever the cause, recognize that the breath is central to anxiety, and restoring it to deep, slow, diaphragmatic breath *through the nose* is one of the most potent pathways out of false anxiety and into a state of relaxation.

POLYVAGAL THEORY

Up to this point, for the sake of simplicity, we've been discussing the nervous system as a dual system with two branches: the parasympathetic (rest, digest, and repair) and sympathetic (fight or flight). In reality, though, like every other aspect of the body, it is far more complex than that, and recent research has proposed an entirely new understanding of the nervous system. In 1994, psychologist Stephen Porges proposed a new model of the nervous system known as polyvagal theory. In this paradigm, our understanding of the sympathetic system remains more or less the same—it corresponds to various states of mobilization (fighting and fleeing). As you'll remember, the sympathetic response is a state of hyperarousal—with the accompanying rush of adrenaline and the consequent rapid heart rate, elevated blood pressure, muscle tension, and shallow breathing. In this state, we might feel anxious, angry, aggressive, and scared.

However, in Porges's model, the parasympathetic system takes on added nuance. Polyvagal theory proposes that humans have two branches of parasympathetic response: ventral vagal and dorsal vagal. The ventral vagal complex is responsible for the functions we typically think of as the parasympathetic response (rest, digestion, and relaxation), while the dorsal vagal complex includes states of immobilization (dissociation, or the freeze response).[5]

The dorsal vagal complex represents a different and, in many ways, later-stage response to stress. Instead of mobilizing, we freeze; this is a state of *hypo*arousal during which we might feel tired and emotionally numb. Dorsal vagal thoughts might include *Everything feels hopeless* or *What's the point?* Being locked into a dorsal vagal response is also frequently associated with past traumatic experiences, where the immobilized and dissociated state was adaptive in enduring and surviving the trauma.

In the ventral vagal response, on the other hand, our body is relaxed; we experience increased lung capacity, we're able to breathe deeply, and we have improved heart rate variability (a measure of the variation in time between each heartbeat, which is associated with health and longevity[6,7]). We have a positive outlook, and our thoughts center around trust, safety, and the sense that we are capable of handling whatever comes our way. In this state, we're able to work through challenges with diplomacy, and we can arrive at mutual understanding. For those who suffer from anxiety, the goal is not only to spend more time in a ventral vagal state (through the practices presented here that cultivate relaxation) but also to carve connections between our dorsal and ventral vagal responses by way of engaging our vagus nerve and reprogramming our automatic responses to stress so that we have a path out of immobilization and hopelessness.

Tend and Befriend

In 2000, psychologists at the University of California, Los Angeles, uncovered that, until recently, we have wholly overlooked another, predominantly female response to stress: *tend and befriend*.[8] "A little-known fact about the fight-or-flight response is that the preponderance of research exploring its parameters has been conducted on males, especially male rats," writes Shelley E. Taylor, professor of psychology at UCLA.[9] In other words, our understanding of the stress response, as with so much of medicine, has left out the unique biology of women.

Thankfully Taylor and her colleagues dove into the research in an attempt to set the record straight. They found that the male response to stress may be more tied to sympathetic arousal—organized and activated by androgens (e.g., testosterone)—and the fight-or-flight theory that has dominated stress research for the last seventy years; the *female* stress response, on the other hand, may be tied, at least in part, to the release of oxytocin and its biobehavioral association with caregiving, otherwise known as the tend-and-befriend instinct. This would make sense, evolutionarily speaking, since, as the researchers point out, a stress response geared toward fight or flight doesn't address the particular challenges faced by females, especially when it comes to protecting their offspring. "The demands of pregnancy, nursing, and infant care render females extremely vulnerable to external threats," the researchers explain. "Should a threat present itself during this time, a mother's attack on a predator or flight could render offspring fatally unprotected. Instead, behaviors that involve getting offspring out of the way, retrieving them from threatening circumstances, calming them down and quieting them, protecting them from further threat, and anticipating protective measures against stressors that are imminent may increase the likelihood of survival of offspring."[10] Instead of fighting or fleeing then, this theory proposes that females, when faced with a threat, tend to their offspring and not only seek out close-knit relationships for protection but may even befriend or ingratiate themselves with those who pose a threat in an effort to keep themselves and their offspring safe. This is a very different response indeed from the fighting and fleeing depicted in our textbooks for decades.

COMPLETING THE STRESS CYCLE

While the fight-or-flight reactions are relatively straightforward, polyvagal theory helps us understand the freeze response. When an animal of prey such as a rabbit is faced with a predator, such as a wolf, its brain does an automatic assessment: *Am I fast enough to run for my life? Am I strong enough to fight back? Or are both of those options utterly hopeless?* In the third scenario, the rabbit suddenly becomes still and plays dead as the predator approaches. This is an involuntary act, and the rabbit becomes immobilized and dissociated from the threatening situation. The predator, in turn, will likely poke around at the limp rabbit, think it's a sick animal, and move on. Once the rabbit's nervous system determines the coast is clear, it will come to and shake vigorously. Shaking is its way of discharging adrenaline and restoring a baseline of relaxation. We humans also face intense stressors, and we are even, at times, immobilized and dissociated in the face of a threat. But the crucial difference between the rabbit and us is that we don't shake. Why not? Well, mainly because we're socially conditioned to avoid doing such things.

Remember the last time you were walking down the street and tripped, just barely recovering your balance before almost falling flat on your face? When that happened, your body experienced a little stress response. Did you take a moment to pause, collect yourself, and shake it off? No, of course not. You continued walking, trying not to draw attention to yourself. But you probably felt shaky for a few minutes afterward as adrenaline continued to pump through your veins. We tend to go through our lives powering through small moments like these as well as more serious threats to our being, such as enduring or witnessing physical violence, living under the oppression of systemic racism, and surviving threats to our existence such as natural disasters and pandemics. Most of the time we allow ourselves scant opportunities to discharge the stress our bodies are experiencing and return to baseline.

When, in our attempt to remain composed, we don't finish the stress cycle, however, our stress never dissipates. Instead, it accumulates. And when we dissociate from or suppress our feelings, our limbic system remains in a state of activation. Subsequently, when the stressor is no longer present, we can't *feel* safe, because the body is still carrying around the patterns of nervous system arousal from those old stresses and traumas. We experience this as anxiety. In the present day, our thoughts and emotions may seem like the issues causing distress, but the real problem is that our limbic system is stuck in the "on" position. Typically, there's no amount of thinking that can flip that switch off—the only way to release it is through reprogramming the nervous system back to a state of calm. This begins with completing the stress cycle.

There are three main ways we can successfully discharge our stress and come back to baseline: movement, self-expression, and connection. Movement can involve dancing, exercising, or even a formal shaking practice. Self-expression can include journaling, singing, playing an instrument, or making art (but this means making art freely, in the manner of a three-year-old; instead of criticizing or saying it's no good, you just draw forth what's alive in you). Feeling connected to others can also complete the stress cycle. This can mean hugging or cuddling; belly laughing or ugly crying; or simply showing up as your raw, authentic self, telling your truth, and having someone really listen and reflect back to you that you're still accepted, you still belong. As Elisabeth Kübler-Ross and David Kessler write in *On Grief and Grieving*: "When someone is telling you their story over and over, they are trying to figure something out."[11] Processing aloud the jumble of emotions we carry and then feeling witnessed and held in this way can be deeply therapeutic to our nervous system.

My favorite practice for completing the stress cycle is shamanic shaking. I learned this in 2012 while studying integrative medicine at the University of Arizona, using a particular track of music called "Amma (Extended Mix)" by James Asher—and that is the music I continue to use today, ten years later. The practice is simple: put on the music, close your eyes, allow a soft bend in your knees, and let your body feel loose like a rag doll. Then simply shake and move in whatever way your body feels like moving for a few minutes.

When your body gets locked, like a frozen computer, into a stress pattern, shaking is like pressing <ctrl-alt-delete>, allowing you to break out of the stress response and return to a relaxation response. Shamanic drums help get your brainwaves into a particularly relaxing pattern known as theta waves,[12] and the shaking movement approximates the way animals shake after a stressful experience. As with anything that completes the stress cycle, this appears to tell the nervous system, in an old and hardwired manner, that the threat has passed, and you are now safe. I also find that the motion loosens up muscular tension and excavates stuck emotions; this can sometimes help unearth unconscious blocks. Occasionally an old memory will surface. When it does, I encourage you to stay with it and meditate on it. Freely shaking has the added benefit of helping you move the way *your body* wants to move, not just the way you think it *should* move. Letting your body call the shots and honoring its needs has the potential to be deeply reparative, reprogramming you to attune to your internal needs instead of outside pressures. I'll admit, it's totally weird; but it's also free, it takes two minutes, and it can go a long way toward completing the stress cycle and lightening your burden of anxiety.

TMJ

Picture a dog about to fight: it tightens its jaw, bares its teeth, and growls. Similarly, during a stress response, the muscles of the human jaw along with the hip flexor and trapezius, which are keenly innervated by sympathetic nerves, tense up. We automatically tense our jaws under stress because, originally, this was a way to signal aggression, strength, and readiness for a brawl. That may be fine for a dog gearing up for a fight, but if you're simply a chronically stressed-out office worker, it's not so great to be woken from sleep with jaw pain. What's more, there's a two-way connection between our jaw and our central nervous system. So, just as stress tells us to clench the jaw, a clenched jaw can communicate back to the brain that we're in a fight, making us feel anxious—and round it goes.

TMJ (which stands for temporomandibular joint) colloquially refers to a common condition, also called TMD, for temporomandibular disorders, of chronic and often painful jaw clenching; this is also frequently connected with bruxism (teeth grinding). There are actually multiple aspects of modern life that can lock us into a pattern of TMJ: (1) unprocessed stress; (2) certain medications and drugs, including some SSRIs and stimulants,[13] as well as illicit drugs such as cocaine and MDMA (3,4-methylenedioxymethamphetamine, also known as ecstasy)[14]; (3) a diet of soft, processed foods (eating real food, especially in childhood, gives our body tactile feedback that helps us develop strong and properly aligned mandibles; as such, when we grow up eating peanut butter and jelly sandwiches instead of gnawing meat off a bone, our mandibles may not form properly);[15] and (4) the neck position typically held while staring at screens and bending to look at phones, which creates tension in the neck and jaw muscle.

Though we can't do much about the PB&J sandwiches from our youth, we *can* complete our present-day stress cycles and take steps to release jaw tension. Interestingly, there is thought to be a connective-tissue relationship between the jaw and the hips. So, if you're struggling to release jaw tension, try some yoga stretches to open the hips, such as pigeon pose, or any pose that helps stretch and release the hip flexor, such as a lunge. Just as a clenched jaw signals that we're about to fight, a relaxed jaw reassures us that we have nothing to fight about.

SITTING IS THE NEW SMOKING— AND EXERCISE IS THE NEW XANAX

You've probably read the headlines touting exercise as an effective treatment for anxiety and warning against the dangers of a sedentary lifestyle. Indeed, nearly every time exercise is put on trial, it is found to be an effective antianxiety agent.[16,17,18] Mechanisms thought to account for this benefit include the impact of exercise on inflammation,[19] norepinephrine modulation[20] (i.e., it decreases stress), and endogenous opioid release[21]—in other words, exercise makes our bodies release homegrown pain relief, which chills us out. Exercise is also an excellent technique for completing the stress cycle.

If you're out of the exercise habit, first of all, I sympathize. Exercise, as a separate activity with its own set of (tight) clothes, can take up a sizable chunk of the day, and in our already over-scheduled and exhausting lives, it can feel hard to fit it in. Maybe you get motivated every year on January 1. You sign up for the gym or buy a package with a personal trainer. Then around January 19, something comes up—you travel for work, catch a cold, or fall out of the habit, and then months go by with nary a push-up.

Well, I'm here to tell you to lower your standards because here's the thing about exercise: it doesn't have to be an all-or-nothing proposition. The truth is that even small amounts of movement significantly reduce anxiety and improve overall energy levels. Somewhere between no exercise and running ultramarathons is your just-right fitness regimen—the movement that feels good for you and can realistically fit into your life.

I used to take multiple ninety-minute yoga classes per week. Between commuting to the gym, changing, and showering, this was at least a two-hour activity from door to door. Now with a busy practice and a family, I don't have two free hours *in a week* let alone in a day. The way I make exercise work in my life is that I do what I call "microcize," which means doing something free, convenient, enjoyable, and quick in my living room or near my apartment for a few minutes. I fit this in right after my daughter's bedtime. Some days I take a quick walk outside; other days I put on Whitney Houston and dance around the living room. Often, I simply roll out my mat and do fifteen minutes of yoga or Pilates. I won't be winning a triathlon anytime soon, but what I'm doing is realistic and sustainable. And when it comes to health and anxiety management, something you can actually *do* is 100 percent better than any loftier but more unrealistic goal. Finding the exercise that works for your life and doing it consistently is Mother Nature's Xanax.

MANAGING PANIC

All the practices in this book create the conditions for less overall anxiety, which increases our capacity to tolerate stressors without dipping into acute panic; but sometimes our anxiety passes the point of no return and we find ourselves in a full-blown panic

attack—a sudden episode of intense fear, accompanied by physical reactions such as rapid heart rate and shortness of breath,
with no real danger or apparent cause. I have seen my patients
enter this territory as they are sitting across from me, flushed and
shaking, asking for guidance. When this happens, I try to help
them do three things: (1) allow the panic to cycle through them,
rather than resist it; (2) become grounded in their bodies; and
(3) become scientists observing their own anxiety.

When we strong-arm anxiety, rather than soften into it and
allow ourselves to feel it, we are actually giving it *more* power.
In his book *Dare*, Barry McDonagh explains how and why we
should let anxiety flow rather than resist it: "Anxiety is nervous
arousal. . . . The secret to recovery is that once you reach a point
where you really allow and accept it, it begins to fall away and
discharge naturally. It's the paradox that is essential to healing
anxiety."[22] McDonagh even advises that we "run toward" anxiety, instead of attempting to run *from* it: "The request for more
is the most empowering and paradoxical move you can make
when facing a panic attack. . . . It's a request anxiety can't deliver. Your fear quickly subsides because the fuel that powers it,
the fear of fear, has been suddenly cut off."[23]

So much of panic is built on our thoughts *about* our uncomfortable sensations or the anxious thoughts themselves, which
create a snowball effect. The emotion at the heart of a panic
attack is not necessarily an insurmountable barrier; it's often relatively manageable and fleeting, capable of being reframed or
challenged (this is where cognitive behavioral therapy excels).
What allows the emotion to spiral is when we attach a narrative
to it, which often stokes the fires of anxiety. When I see a patient
panicking, I do my best to model acceptance and calm in the
face of it. Rather than reacting in fear, I attempt to show them

that *I* can handle their anxiety, in order to help them see that *they* can too. Only when we let the panic flow can it come to full resolution.

In the absence of a trusted friend or therapist to help you through an acute moment of distress, another way to ride the wave and get to the other side with more ease is to physically ground yourself in the present—in essence, to remind yourself that you are still alive, still breathing. Panic is like a runaway train, or like energy spiraling upward out of our control. Splashing cold water on your face or opening a window for a blast of fresh air can help bring you back into your physical body and into the present moment. Certain yoga poses can also be helpful. I like child's pose best for these moments. First sit on your knees, and then fold forward to rest your forehead on the floor. Let your arms fall comfortably down by your sides. If you're somewhere that you can't exactly drop to your knees (say, an office with an open floor plan or airport security), simply sit and focus on feeling your body supported by the chair. Another effective grounding technique is to count five things you can see, four things you can hear, three things you can touch, two things you can smell, and one thing you can taste. This trains your attention on the present moment. Panic is often borne out of "future tripping" or dwelling on the past, where we tangle with imagined problems or unchangeable grievances. Present moment awareness is like garlic to the panic vampire. Once you are back in your body, remind yourself that you are experiencing panic—and that this is only a stress response. It's exquisitely uncomfortable, but you are safe.

Finally, I urge my patients to explore their panic with the dispassionate curiosity of a researcher. Take inventory of the sensations in your body: *Heart is pounding; breathing, rapid; hands, shaking.* Think to yourself: *Isn't that interesting: This is my body in a stress response. I know this now. That is all this is. Look how well my body works, doing what it's supposed to do when I am feeling anxious.* This shift in perspective—seeing panic as an indication that your body is functioning properly, rather than an indication that something is going wrong—can be very helpful. It takes the emotional power out of the reaction, reframing it instead with interest, curiosity, and even appreciation, rather than fear.

Panic-Attack Cheat Sheet

If you experience panic attacks regularly, it can be helpful to have a handful of reliable strategies to help you through these difficult moments. The list below offers some ideas for fast and effective interventions. I recommend writing a few of these down and keeping the list in your wallet or on the fridge.

- Go outside and move your body to release accumulated adrenaline.
- Shake to shamanic drum music—it works for completing a stress cycle as well as panic.
- Refocus on sensory information in the present moment:
  - Count five things you see.
  - Do a 4-7-8 breath.
  - Count four things you're touching (e.g., legs, sweater, floor, and chair).
  - Do a 4-7-8 breath.
  - Count three things you can hear.
  - Do a 4-7-8 breath.

- Count two things you can smell.
- Do a 4-7-8 breath.
- Count one thing you can taste.
- Take box breaths—inhale for 4 counts, hold for 4, exhale for 4, hold for 4, and repeat.
- Ground your feet and push into a wall with both hands.
- Count backward from one hundred by sevens.
- Run your hands or feet through something sensory, such as water, sand, or Play-Doh.

Flight Anxiety

More than half of Americans experience anxiety about flying. Some of it can be attributed to false anxiety generated by the typical stressors of air travel: sleep disruptions, rushing through the airport, worry about missing a flight, TSA procedures, missed meals, and eating fast food. But I also think about flight anxiety in terms of Ayurvedic medicine's approach to *doshas*, or bioenergetic types.

In Ayurveda, an ancient healing system from the Indian subcontinent, there are three main doshas: *vata*, *pitta*, and *kapha*. Vata, in particular, is characterized by a tendency toward cold, dryness, motion, and change as well as racing thoughts, worry, restlessness, and anxiety. Vata is also the air element, governing movement. The best way to keep vata in a state of balance is with the consistency of a daily routine. If you think about air travel, it's the perfect storm of vata exacerbation—it disrupts our routine, and it involves hurtling through the air in a cold, dry plane. There is nothing quite like changing time zones and *flying through the literal air* to get vata especially off balance. This may be why vata types, as many of my anxious patients are, find themselves feeling like a ball of anxiety on flights, throwing back Xanax and clutching the armrest every time there's turbulence.

The best antidote to flight anxiety, then, is to balance vata. On a travel day, this balancing includes staying warm by wearing a scarf and cozy socks; packing herbal tea bags in your carry-on so you can sip something like warm tulsi tea on the plane; eating regular meals (no skipping breakfast) and favoring warm foods cooked with healthy fats while avoiding raw, cold foods; getting plenty of sleep (which sometimes requires paying more to avoid a 6:00 a.m. or red-eye flight); avoiding stimulants, such as coffee and sugar; and, if you like, doing a calming aromatherapy ritual with vata-soothing scents, such as bergamot, sandalwood, and rose. All these practices will help keep your anxiety level lower during travel. And at the end of the day, the most important thing is to be patient with yourself, recognizing that travel days can be an anxiety vortex and you will automatically feel calmer once you're settled at your destination.

Finally, I suggest bowing to the metaphorical power of flight, in which we are hurtling through space with very little control indeed. In other words, it's a lot like life. Control was always an illusion; we never had it. Begin to see flight anxiety as a form of true anxiety, trying to tell you to let go of your need for control. We're not the ones flying this rock. What if we surrender and trust whatever is, knowing we will ultimately arrive at our destination? It's nice not to be the one steering sometimes.

PART III

True Anxiety

Chapter 12

Tuning In

How to stay connected to your soul:
When something happens in the world that
is wrong, don't try to move on with your life
like it is right. The voice within you that
says, "this is not okay" is a direct call from
the basic goodness of your spirit. Pick it up.
Every time. Pick it up. And stay on the line
until you figure out how to help.

—*Cleo Wade*

Sometimes you can fine-tune and optimize every aspect of your physiology, but you're still left with discomfort or an inability to relax or feel optimistic about your life. This is true anxiety, which we can think of as an emotional compass telling us, *Something is not OK*. These sensations and emotions aren't something we should attempt to eradicate, nor could we if we tried; they pertain to our insights, our traumas, and our deepest sense of vulnerability and purpose. When we understand that our anxiety has important information to offer, a crucial shift occurs. Our uneasy feelings are no longer the enemy or something to vanquish—they become our tools and allies instead.

That said, the insights offered by our true anxiety aren't always transformational on such a grand scale. On any given day, your inner compass might be leading you directly to your destiny, or it might simply be guiding you toward the next right step, which could be as simple as having more patience with your child or giving yourself a day of rest when you need it. But even these small threads in the tapestry of your life are infinitely impactful. True anxiety is here to give you a little nudge and tell you that it's time to *leave that thankless job* or *set some boundaries in that relationship that isn't serving you* or *create something and make your unique and powerful offering to this world*.

We are isolated and lonely, overworked and worried, estranged from nature and exhausted, each of us on our own separate hamster wheel, disconnected from community and creativity, some of us hermetically sealed off from the suffering around us, others drowning in it. If the world doesn't feel safe to you right now, it might be because it isn't. Violence and discrimination abound. If you are a member of any number of vulnerable and marginalized populations in our society, there is good reason for you to be anxious. In fact, our ability to change as a society depends on you listening to your anxiety, and society listening to *you*. This is a moment of much-needed reckoning. And it is time to confront the powerful truths our anxiety offers.

In order to hear that truth, however, we have to get quiet—and most of us will do just about anything to avoid stillness. When was the last time you stood in line for the bathroom or waited for the elevator and didn't pull out your phone? We often think that we're being productive by checking our phones while we have nothing else to do when, in fact, it is just this kind of moment—one of simply being with ourselves and our thoughts—that is necessary to hear the quiet whisper of our true anxiety. Being able to connect to the truth of what it might be telling us requires stillness and silence as well as being ready, willing, and able to surf whatever emotional waves arise.

We find this state of quiet acceptance difficult for a few fundamental reasons: first, we're taught from a young age that when something is difficult, it is necessary to distract ourselves. When a child has a tantrum, we think, *How can I make the crying stop?* We know that if we hand the kid some sugar or a screen, they'll probably be satisfied. Problem solved, right? Well, actually, now we've taught this kid: I can't handle your big emotions, *you* can't handle your big emotions, and should you ever feel big emotions in your future life, quickly find something that will distract you, offer you a hit of dopamine, or numb you out. It's no wonder even we adults turn to our phones or emotional eating when in fact we just need to feel our feelings and let our tantrums run their course.

But we also struggle to sit with ourselves in silence because we live in an era of climate control and instant gratification. Let our rooms be forever 70 degrees Fahrenheit, never a degree hotter or colder. Real life, however, is not climate controlled, and we cannot uniformly medicate ourselves out of sadness, insomnia, or distraction. We are being cognitively restructured by such impractical promises. We feel entitled to hear any song, see any movie, or pull up any fantasy with a few keystrokes; we can see clear across the world through FaceTime; we can tamp down deliberate reflection by scrolling through TikTok. But to be available for the truth of our anxiety, we need to be available for discomfort. Sometimes our truth is a blizzard, and we need to be willing to sit in the middle of the storm for a period of time to take it in and enact its wisdom. And leveraging that anxiety to fuel meaningful change can sometimes feel as gradual as carving canyons with rivers.

Perhaps most crucially, we're almost always surrounded by influences that suck us out of the present moment. Even when we're not having a tantrum and being appeased by a screen or a snack, we are constantly being seduced by other distractions, pursuing everything from better shoes to a better body to a better house. And humming beneath it all, of course, is the implied assurance of escaping our mortality.

True anxiety, though, is not a nuisance or a debilitating symptom to be suppressed with meds or ignored in exchange for shiny promises. In order to hear it, we need to slow down, get still, and listen more closely. And the only person who can hear its stirring truth is you.

THE BODY WHISPERER

To listen is to lean in, softly, with a willingness
to be changed by what we hear.

—*Mark Nepo*

Though true anxiety usually starts as a whisper, if you don't listen to it, it will develop over time into a shout. And usually, it is the body that communicates on anxiety's behalf when we have not paused for long enough to hear what it has to say. This is the case with both false *and* true anxiety. For example, if your false anxiety is related to blood sugar instability, you might start out with mild dysglycemia, meaning you might feel a little "hangry" and uneasy once in a while. That is your body whispering. But as these symptoms worsen, your anxiety becomes stronger and more consistent. Now, when you hit your 5:00 p.m. blood sugar crash, you might experience panic. That's your body shouting, *Help me, I have an unmet need!* The difference with false anxiety is that the solution is swifter and more straightforward: keep your blood sugar stable.

And yet even if the solution tends to be a bit more intricate with true anxiety, a similar process occurs. If you're in the wrong job or the wrong relationship, your body might start by whispering, *Something doesn't feel right.* You might start to notice a vague feeling of discomfort or restlessness—perhaps the way someone spoke to you isn't sitting right, or you feel off-kilter in a meeting. It's easy enough to brush off these feelings in the moment, but if you systematically ignore these warnings, your body will eventually raise its voice to gain full command of your attention. When it reaches the point where you can't get out of bed, or you have repeated panic attacks at work, or you shut down when you're trying to be intimate with your partner—this is the body shouting. The body uses these symptoms to assert itself, communicating, *I refuse to continue under these circumstances.* This feels more like a spiritual crisis. Of course, it's also possible to feel that you're having a spiritual crisis when it's really just a blood sugar crash, but there is usually some predictability to the timing of false anxiety—you tend to spiral after a sugary coffee drink, or you feel hopeless desperation only when you're sleep deprived. With true anxiety, there isn't the same regularity. There is, however, a consistent *theme*. Listen to the language of your anxiety. Do you panic on planes or elevators? Do you feel most anxious when you're alone? In crowds? When your partner gets home from work? The themes of your anxiety offer clues to unconscious issues at play. If your panic centers on being alone, it may well be telling you to reclaim community in your life. Perhaps your friendships feel lonely. This might mean you need to show up as your authentic self or find different friends. If you panic in elevators and the theme seems to be centered around being stuck, ask yourself where else in your life you feel mired. In your job or a relationship? Do you feel stranded saying yes to every request and putting everybody else's needs before your own? If this is the metaphor of your anxiety, then you would do well to start speaking up for your needs and working toward freeing yourself.

The messages sent through the body by anxiety can often lead you straight to the heart of the matter. After all, "there is more wisdom in your body than in your deepest philosophy," as Friedrich Nietzsche wrote. If you fearlessly inquire, and remain still long enough, you *will* eventually discern what your body is trying to say. At that point, your job is to trust what you hear. People fear that if they venture into the dark corners of their feelings, they'll never come out. But, in fact, it's the opposite. The less we try to strong-arm and resist anxiety, the more easily we will flow with it *and* flow out of it.

Some of us believe that we can get away with ignoring the body's messages because it can't speak in words. We think it's OK to betray ourselves as long as we're pleasing everyone else. But, ultimately, our bodies call us out. The body witnesses all, and it's a persistent communicator—it *will* eventually make itself heard. So, if you're finding it difficult to face up to your anxiety, remember that confronting the potentially inconvenient truth now could save you from a barrage of inconvenient symptoms over a lifetime.

Permission to Feel

Once, when I was standing in the back of a synagogue at a funeral, allowing the waves of grief to overtake me, someone leaned over and whispered, "Be strong, don't cry." We live in an emotion-phobic culture that predominantly values emotional evenness and stoicism. Vulnerability and sensitivity are seen as signs of weakness. When we cry, we *apologize*, instinctively holding it back, rather than letting it flow. Every once in a while, we have to step back and ask, *How is that working out for us?* Judging by our skyrocketing rates of anxiety and depression, I think we got this wrong. It's time for us not only to give ourselves permission to feel but also to reconceive what it means to do so. It is far braver—not to mention healthier—to dive into our difficult emotions as they arise than to suppress or ignore them.

In the mental health professions, we often summon psychoanalyst and founder of analytical psychology Carl Jung (1875–1961) by offering his insight that what we resist persists. That is, no emotion has ever been successfully pushed under the rug. When we think, *I don't want to feel that; I'm going to force it out of my awareness*, the feeling doesn't go away. If anything, it actually doubles down and gets lodged, frequently transmuting into chronic back pain, or headaches, or digestive issues, until it eventually reaches a boiling point that prompts us to lash out. So, if a tidal wave of emotions is coming at you, try diving into it instead, letting it overtake you. Feel the full strength of your sadness or rage or grief. Just as with waves, these emotions eventually crest and resolve. You are far more likely to come out the other side feeling OK if you flow with the wave of emotion, rather than try to outrun it. As psychologist Marc Brackett, PhD, writes in his book *Permission to Feel*, "If we can learn to identify, express, and harness our feelings, even the most challenging ones, we can use those emotions to help us create positive, satisfying lives."[1] In other words, what the person should have said to me at the funeral was, *Be strong, cry.*

PRESENT MOMENT AWARENESS AND MORTALITY

My patient Jada came to every one of our sessions with a pen and a notebook; she took extensive notes on what we talked about and the recommendations we discussed. This is not an unusual occurrence, actually, especially for patients with anxiety. Often they are striving for control, working to wrestle life into shape, and they are not going to let anything slip through the cracks. So, I wasn't surprised when Jada told me that she was taking charge of her dad's health after he was diagnosed with a fairly serious heart condition. Jada described the possibility of her father's death as the "worst-case scenario," and often stopped midsentence, saying she couldn't bear to think about it. Her instinct was to govern the situation as a way of drowning out these intense feelings. She called him often, though she no longer asked, "How are you?" She boxed out any chance for her father to process what he was going through, instead spending the time admonishing him for drinking soda, urging him to exercise, and escalating his search for the "best doctors." I pointed out to Jada that he might feel castigated at a vulnerable moment in his life and that she was seeing him as a problem to solve as opposed to *seeing him*. A layer of fear and distance became firmly wedged within their once natural connection.

Here's the thing: when the "worst-case scenario" eventually does happen, the best way to survive it is to surrender ourselves fully to the experience. There is no way to commandeer fate and no avoiding mortality, for us or our loved ones. Everything we hold dear, we will one day lose. That *is* true anxiety; this is the distilled fear at the center of it. If we are to live fully, we must lean into the exquisite experience of being human, with all its vulnerabilities and emotions—joy, rage, devastation, grief.

I counseled Jada that the only way for her to gain some peace in this anxiety-inducing situation was to embrace the circumstances as they stood. Our unconscious prefers that we numb or distract ourselves in the face of impending pain. But in avoiding vulnerability, we can also miss the raw experience of what makes our lives meaningful. It's better to remain wide awake. I suggested that Jada offer her advice to her father and then step back and practice radical acceptance—that is, the complete acknowledgment of life in the present moment. For Jada this meant accepting the reality that one day her father would die. But right now, right here, she was able to show up for him, to be with him wholeheartedly—and feel the gift, and poignant anguish, of loving someone so much.

BICEP CURLS FOR PRESENT MOMENT AWARENESS

Meditation is both a path out of anxiety[2,3] *and* a tool for listening to it. A cardinal feature of anxiety is "future tripping"—in other words, worrying about the future. Meanwhile, a primary goal of mindfulness meditation is cultivating present moment awareness, which has been found to limit future-focused thinking.[4] In fact, a growing body of research suggests that mindfulness meditation appears to coincide with changes in the brain that enhance mood regulation—in other words, strengthening our ability to remain present appears to decrease anxiety.[5]

To me, meditation is not a skill but an act of showing up, routinely, and sitting still. You don't need the perfect moment after you've gotten to the end of your to-do list—because, of course, that moment will never actually come. All you really need is to sit down and remain in stillness and silence for a few minutes. Attempt to hold your attention on the experience of breathing; feeling the inhales and the exhales, what it is to be alive in your body at this moment. Approximately one nanosecond after you've begun, a thought will pop up. With few exceptions, this thought will pertain to the future or the past. You'll catch yourself thinking about your grocery list and that thing someone said ten years ago that's still irking you. Acknowledge the thought, and let it pass. A nanosecond later, another thought will pop in. Let it pass. Each time a thought comes, picture yourself sitting just a bit to the side of yourself, observing your mind as it goes through this process. "You are not the voice of the mind," Michael A. Singer writes in his book *The Untethered Soul*, "you are the one who hears it."[6] As thoughts continue to arise, maintain an attitude of patient bemusement, like a grown-up watching a child struggle to learn to use a fork, barely able to get food from the plate to the mouth. With each of these thoughts, pull your attention back to the present—back to your breath.

Many of my patients tell me they don't like meditating because they're not good at it—meaning, their minds wander too often. So, let's correct the public record here: your mind *will* wander during meditation. That's not a failure; that's the gig. There is no such thing as being bad at meditating. Meditation is simply about showing up and giving that muscle of present moment awareness a workout. Each time your mind wanders is an *opportunity* to strengthen that very atrophied muscle. And each time we pull our attention back to the present, we're doing a little bicep curl. Before long, we're going through life with a toned muscle for catching ourselves and choosing how to respond before falling into our habitual reactions.

A common misconception about meditation practice is that the primary goal is to become blissed out—that once you start meditating, life is good vibes only. This notion misses the point; the world is not in the slightest bit "all good"—it is teeming with suffering and injustice. Meditation is useful for helping us connect to the anger, grief, and sadness that arise as a result of all this suffering and injustice. For me the ultimate goal of meditation is to reach the unadulterated truth. In my meditation practice, I try to start with a neutral mindset, but if negative thoughts come up, I stay with them, particularly as I have found that they reliably lead toward my true anxiety. This helps me let what's wrong in the world set my trajectory for action in my waking life.

See if you can approach meditation as an open-ended question, an invitation for truth. Then, when you least expect it, you will have a moment or two when you're actually present—when you're not simply watching the movie of your thoughts but are in the experience of the present moment. Amid this glorious nectar, your unconscious will find it safe to tiptoe forward with a kernel of truth. It may be subtle, or it may hit you like a ton of bricks. Either way, when you receive a download straight from your intuition, don't question it or overthink it. Just hear it. Transmissions like this can make you feel lighter, like something finally clicked into place, or they can be heavy, like a painful awareness surfacing after so long. Either way, this missive is an essential part of who you are.

Chaos-Seeking in Adulthood

Some people grow up in chaotic homes. By "chaotic," I don't just mean hectic, but, rather, households where kids either don't have consistent, reliable caregivers or cannot necessarily count on getting their basic needs met. Growing up in this kind of environment establishes a baseline of chaos and disorder, which, in addition to being a precursor to anxiety, can make it difficult to be comfortable with stillness in adulthood. Adults who grew up in chaotic homes may forever find themselves filling their lives—and any potential for quiet—with stimulation, because it feels familiar and therefore comfortable. And when the chaotic home was also traumatic, anxiety can become a form of escape. That is, anytime the adult child of a chaotic home finds themselves in a state of tranquility or silence, trauma can creep back in. So, unconsciously, they're driven to saturate any silence with anxiety—a kind of frenzied state of distraction that can eclipse the traumatic memory tapping them on the shoulder. In these circumstances, *the anxiety itself* serves the unconscious need to avoid stillness. This can be very difficult to unwind, because the mind will powerfully resist stillness, but the salve is to persist and increase your capacity for sitting quietly, acknowledging that it's especially difficult for you and being patient and compassionate with yourself as the urge to find a distraction arises. Make sure you have a supportive container in place (e.g., regular therapy with a trauma-informed therapist) should traumatic memories arise. Stay gentle, but also stay the course, reminding yourself that there is guidance in the true anxiety that will arise in these hard-won moments of stillness.

The beautiful, life-changing thing about meditation is that, when you practice on a regular basis, it starts to permeate your whole life. You walk down the street more mindfully. You interact with strangers more mindfully. And you dance with your anxiety more mindfully. An *Oh, God, I can't handle this* spiral transforms into *OK, I can see that this situation is happening right now, and it makes me feel really anxious.* You become less *identified* with your thoughts and instead get accustomed to being the *observer* of your thoughts. You begin to realize that these thoughts almost always have a quality of anticipating some bad outcome in the future and that these are *not* the wise soothsayers of true anxiety. When an anxious thought bubbles up, mindfulness gives you a bit of distance. This is the fundamental power of strengthening that muscle of present moment awareness: carving out a moment between stimulus and reaction, so we can choose how to *respond* with intention, instead of defaulting to a familiar emotional *reaction.* When I allow myself this conscious pause, I'm able to hew more closely to compassion and understanding, which offers a more peaceful ride through the challenges of our lives. But be careful to manage your expectations: for me, I manage to pause and catch myself maybe 10 percent of the time. (If there are extenuating circumstances, say you're living in quarantine during a pandemic with your in-laws, aim for a 2 percent success rate.) Anytime you succeed, however, take a moment to notice your progress. How you respond to yourself in moments of "failure" is equally important. Rather than berate yourself—which is often just a perpetuation of the conditioning we received from our parents, who in turn were conditioned by *their* parents—support yourself for trying. Offer yourself compassion for how very challenging it is to stay true to this intention in daily life. And when things get really juicy in your meditation practice, you will start to recognize that there is a great deal of calm in the present moment. As Eckhart Tolle points out, "Your life situation may be full of problems—most life situations are—but find out if you have any problem at this moment. Not tomorrow or in ten minutes, but now. Do you have a problem now?"[7]

Gratitude Practice

In my experience with patients who suffer from anxiety, focusing on gratitude reliably shifts their moods and expands their outlook. A gratitude practice is a rebellious act, meant to be done in imperfect conditions. We are berated daily by messaging that tells us we're not enough. Gratitude recognizes a deeper truth: that we can find abundance even during our darkest moments. This does not invalidate the very real causes of fear and suffering in our lives. But our brains are, by design, wired to dwell on this lack—it's an instinct that helps us survive. We can even be *grateful* for this tendency, as it kept our ancestors alive. However, we don't always need to give it quite so much airtime.

I suggest keeping your gratitude practice simple: write down or say out loud three things you're grateful for each day. That's it. No matter how not OK things might feel (and be), a gratitude practice forces you to focus, however momentarily, on the things that *are* OK. Over time, your brain will forge new neural pathways[8]—including modulation in the medial prefrontal cortex, which is involved in resilience and mood regulation. As this occurs, you may notice a broadening of perspective from a myopic focus on what's wrong to the larger and more complex truth of what *is*.

TRUE YES, TRUE NO

The concept of "true yes" and "true no" comes from *Nonviolent Communication*, a book, school of thought, and training course by the late psychologist Marshall B. Rosenberg. His teachings help people compassionately identify and satisfy the unmet needs within them in order to bolster their relationships with others. Navigating the world in this way prevents us from betraying ourselves; sometimes true anxiety taps us on the shoulder because we have, in some sense, abandoned ourselves. A critical part of this work is discerning when you can offer a true *yes* or a true *no* when someone asks something of you. It may sound simple, but you would be surprised how often we have trouble actually recognizing when we deeply do *not* want to comply with a request, and acting accordingly; for many of us, it goes against the grain of our personas. We so keenly want to be liked or avoid confrontation or make the wrongs of the world right that we put these instincts above our own self-preservation. But, in the end, depression and anxiety, as Rosenberg pointed out, are often "the reward we get for being good." If we are to find a harmonious relationship with our anxiety as well as hear the truth buried within it, we need to become well versed in listening for our body's true *yes* and *no*. This is, in fact, another language that the body speaks—one that is expressed in sensations that are on a spectrum between contraction and expansion.

The choreographer and mother of modern dance Martha Graham observed that every movement is either a contraction or a release; that is the language of the body. You can feel contraction or expansion in your muscles, your diaphragm, and your breath. A true *no* occurs when you consider the idea of something you've been presented with and you know, physically, that it is not the right choice for you. It feels like a contraction, clenching, tightness of breath, or perhaps a knot in your stomach; it can feel cold; it's the feeling of *ack, ugh, yikes, nope.*

A true *yes,* on the other hand, elicits warmth, release, expansion, openness, and ease in the body. It can feel like a weight being lifted, a tension releasing, or a positive sensation of butterflies in your stomach; it's a feeling of the body saying, *That sounds right* or *I'd like that, yes.* For me there's a particular tingling sensation that tells me something salient is happening, as if my next move has just been revealed. A bodily *yes* goes beyond feeling drawn to pleasure—sometimes it's a deeply felt sense of alignment with a difficult but important duty. These truths we feel in our body can be mundane or profound—what matters most is that our words and actions correspond with the feeling coming from deep within.

Most of us fall into the trap of the false *yes* when we can't hear, or simply ignore, our bodies' directives. This leads to the regrettable favor we committed ourselves to doing or the too-low salary accepted in a negotiation or the unwanted touch that we talked ourselves into allowing. Can you remember an interaction like this and call up that sensation right now, how it felt in your body? Have you ever noticed your gut saying, *Hey, slow down before you say something you know isn't true,* but you steamrolled right over this warning? Finding your *no* starts by becoming more measured in making day-to-day decisions, by taking a minute to check in and listen before moving forward. Of course, it's sometimes necessary to agree to something for the sake of healthy compromise, to make a reasonable concession over a gut feeling—if we have to pay our dues, for example, and agree to an assignment at work that might really feel like a *no.* But stay awake to the fact that every false *yes* is a little betrayal of self—and, ultimately, this is how we train our bodies to be confused, or quieted, about our essential truths.

And remember, when we give our false *yes*, it *never* ends well: people count on us, but we flake at the last minute; or we go through with whatever it was we promised, and in so doing, we stretch ourselves too thin; or we end up resenting the other person. Ask yourself: Would you want someone to tell you *yes* but later resent you? The surprise twist is that it often ends up being more compassionate to tell the truth in the first place, rather than reflexively agree to whatever is asked of us. Not being straight with your colleague, family member, or friend is unkind to them—and saying yes to keep the peace is unkind to yourself.

I do "true yes" and "true no" work with many of my patients. For Yelena, my thirty-three-year-old patient from Kansas, it was transformative. Yelena is a caretaker extraordinaire. She grew up looking after her three younger brothers, and today she is a social worker. She also has a number of friendships in which people rely on her far more than she relies on them. Perhaps it will not come as a surprise then, given the amount of other people's stress Yelena shoulders, that she was operating with a compromised ability to perceive what her body was telling her. When Yelena first came to see me, she struggled mightily with focus and anxiety, and she relied on Xanax to manage her panic attacks. Over the course of several years, we worked on her false anxiety, getting her less inflamed and better nourished. Once we finally got Yelena to the point that she was no longer held back by physical anxiety, we reached a deeper rhythm in therapy. She was ready and eager to connect with herself—to feel her way to her own truth—and she responded to true *yes* or true *no* particularly well. In fact, she became so attuned to the physical sensation of the true *yes* in her body—in the form of excitement and lightness—that it became the internal compass that guided, and revolutionized, her life. She changed jobs, restructured her friendships, moved across the country, got in (and eventually out of) a romantic relationship, and checked into a detox clinic to come off Xanax. This last effort was actually against my recommendation. I thought she should taper off Xanax more gradually than the detox program would have her do, but she was determined that this was the right next move for her. Every true *yes* decision she'd made up to this point had been appropriate and wise, so I deferred to Yelena's discriminating judgment. Since her instincts had taken her this far, this triumphantly, I didn't feel I could rightly claim to know better. As it turned out, the detox program was a good fit, and she came off the Xanax successfully.

But not all of us have Yelena's capacity to cultivate such intuition. It would be easier to navigate, of course, if our true *yeses* and true *nos* more often aligned with what is expected of us in our culture. The trickiest part of this process is when the world is telling us we should do the thing that makes our body contract. Meanwhile, the thing that gives you a resounding *yes* feeling in your body is sometimes exactly what everyone around you is telling you is impractical, unrealistic, *not gonna happen*. These societal and familial expectations can tempt us into betraying our true *yeses* and *nos*, or worse, they can throw us off course, persuading us that societal instincts are actually our own.

Women, in particular, have been conditioned to say *yes*—in an effort to people-please and meet the needs of all those around us—without regard for our own personal truth or energetic limits, for as long as history. "By the time most of us get to adulthood—especially those of us who come from a history of societal oppression—we have lost all capacity to listen to ourselves, to trust ourselves," Holly Whitaker astutely points out in her book *Quit Like a Woman*. "We seek answers outside ourselves because we are told time and again that our innermost intelligence is wrong."[9]

Many of us experience a distinct challenge in differentiating between fear and intuition. The two feelings are often so inextricably woven, it can take real effort and discipline to learn how to untangle them. I agree with Glennon Doyle that the difference between the two can be likened to a vibrational frequency—fear registers as a higher-frequency wavelength, almost like a trill, and intuition is a slow, longer wavelength.[10] I myself have struggled with distinguishing between the two for most of my life. Until recently, I thought I should play up the rational and objective side of myself in order to be taken seriously, to be accepted in the boys' club, to have power. Despite the fact that I had a strong sense of intuition, I tamped all this down, for fear of being labeled irrational.

Then I started studying alternative healing modalities. With this, my intuition became an irrepressible force, and I witnessed firsthand how powerful holistic therapies can be. Once I began to peel away the layers of indoctrination, I began to realize that I had been denying a range of potent skills within my nature. My true anxiety, my true *nos* and *yeses*, my intuition—these all make up an inner compass that had been deeply buried; it now serves me favorably in my life *and* in my profession. Which is not to say I've turned my back on rationality and objectivity. I do my best to bring both sides of myself—the analytical and the mystical—to my decision-making process. I review the data and think through the practical advantages and disadvantages, but I also listen to what my body and intuition are telling me to do. In the end, my own well-being and purpose depend on a constant discernment—between a true *no* and a false *yes*, between intuition and fear, between societal conditioning and my own inner knowing.

How to Say a True No

So many of my anxious patients are people-pleasers. This is understandable, as many of us have endured a lifetime of conditioning that teaches us that the world expects us to comply with any and all requests. Sometimes our childhood environment taught us that our compliance wins parental approval or keeps the whole household afloat. And the resulting scramble to satisfy everyone leaves us harried, disconnected from our own needs, and anxious. For recovering people-pleasers, it can take some practice learning how to state your truth respectfully and move forward. Here are some gentle ways to honestly, firmly, and diplomatically express a true *no*:

"Thank you for the invitation. I'm doing my best to prioritize my family and work these days, so I'm minimizing social engagements for the time being."

"I am so honored that you thought of me for this project, but I don't have the bandwidth at the moment."

Or, in the words of Melissa Urban, cofounder and CEO of Whole30 and armchair boundaries expert:

"What you are asking me to do makes me uncomfortable. Is there another way I can help?"[11]

ALLOWING FOR INTUITION IN THERAPY

One of the most stunning points of Holly Whitaker's book *Quit Like a Woman* calls attention to the fact that Alcoholics Anonymous, created in the 1930s by upper-middle-class white men, offers guidelines—known as the Twelve Steps—that don't fundamentally address women's needs. "To be reminded you're not God, to become right-sized, to refrain from questioning rules, to humble yourself, to admit your weakness, to chronicle what's wrong about you, to be vulnerable enough to admit your faults to another person, to shut up and listen: these are all behaviors associated with (and imposed on) women," she writes. "They are in essence instructions on how to be a woman, and to those men, they were medicine. To act in this manner was a crazy, new way of being, and felt like freedom. But to a woman or any other oppressed group, being told to renounce power, voice, authority, and desire is just more of the same shit. It's what made us sick in the first place."[12] Amen.

When it comes to reconnecting us to our inner knowing, I believe that cognitive behavioral therapy, or CBT, is worth reexamining through the same lens. CBT, pioneered by psychologists such as Albert Ellis in the 1950s and Aaron T. Beck in the 1960s, is one of the most popular forms of therapy in America today—often used to treat anxiety. It guides patients to understand that their thoughts and emotions can be faulty sources of information and that our personal history informs how we interpret the world, a phenomenon known as cognitive distortion. When CBT is helpful, it allows patients to recognize and change destructive thinking patterns. My patient Marcus, for instance, who had long experienced panic attacks in airport security lines, was able to stop this from happening entirely with a CBT workbook. Indeed, I use some CBT techniques in my own practice. Our thoughts *are* influential, and they can sometimes cloud our perception. But I *also* believe that our thoughts and emotions can be powerful sources of information. And in that respect, I believe that CBT, with its underlying suggestion that we should question our feelings, can sometimes be undermining—particularly for women.

CBT broadly says emotional reasoning is not to be trusted. Patients might describe a feeling of being excluded or disliked in a social interaction, and CBT suggests that they shouldn't believe their own thoughts or feelings—these dark hunches are simply an example of a cognitive distortion, such as mental filtering, discounting the positive, mind reading, or catastrophizing. *Don't be so emotional*, CBT says. *Be objective.* In short, this cognitive therapy celebrates what, broadly speaking, comes more naturally to men and discounts what is, broadly speaking, second nature to most women. *I know, I know*, I believe in the exceptions, too: so much of what we understand as gender is a false and socially conditioned construct; not all men think objectively, and not all women are emotional; and CBT can be incredibly helpful and insightful for men and women alike. But, stepping back, I also believe that this notion that dispassionate thinking is more valuable than feelings and intuition exists at the heart of CBT—and it's potentially harmful.

Humans are sophisticated social creatures. We pick up on so many data points: subtle facial expressions; physical gestures; when a person laughs or lets a joke flop. Yes, we have biases, and we can come away from an interaction with the wrong impression, but we also have refined machinery for assessing how others feel—women, perhaps preternaturally so.[13] And dismissing these hunches effaces a person's hard-won sense of reality. I consider my patients' views as something to explore, whether accurate or not (we will likely never know). Our feelings offer evidence—not necessarily the whole story, but a useful piece of it. Feelings aren't facts, but they're not hysterical falsehoods either. They are a form of truth.

Bear with me while I take you on a (relevant, I promise) detour. In recent years, children's cartoons have taken a step in the right direction in their depiction of young heroines. No longer damsels in distress (or literally requiring a kiss from a prince to regain *her voice*, as is the case for Ariel), today's young female protagonists are fierce, as much warriors as the boys—think Astrid in *How to Train Your Dragon* or the eponymous heroine in *Mulan*. But this shift only takes us part of the way toward a representation of true female power. The message of these movies is *Girls are as powerful as boys when they take on characteristically male traits and hobbies*. With notable exceptions—such as Moana and Raya, who use such characteristically female skills as emotional attunement, trust, and collaboration to save the day—few of these influential characters represent the true superpowers of women. I have nothing against girls dazzling everyone with their battlefield courage and skillful swordplay, as long as we recognize that it's also OK to channel that ferocity into whatever your female heart desires, be it playing with dolls as a way of mastering complex interpersonal dynamics or dressing up in a fashion that makes you feel fully expressed. As parents, we respond to the cultural sea change by attempting to steer our daughters toward sports and the STEM fields of science, technology, engineering, and mathematics instead of dolls and dress-up, and while sports and STEM are wonderful interests for boys and girls alike, this sends a message to our daughters that what they might have been otherwise inclined toward is somehow inferior. And this further reinforces the presiding cultural consensus: that what males are phenotypically inclined toward is inherently more valuable.

Actress and filmmaker Brit Marling wrote convincingly about this issue in a *New York Times* op-ed, sharing that "the more I acted the Strong Female Lead, the more I became aware of the narrow specificity of the characters' strengths—physical prowess, linear ambition, focused rationality. Masculine modalities of power. . . . It's difficult for us to imagine femininity itself—empathy, vulnerability, listening—as strong. When I look at the world our stories have helped us envision and then erect, these are the very qualities that have been vanquished in favor of an overwrought masculinity."[14]

We may have a long way to go as a culture before we can fully value characteristically masculine and feminine traits equally, but my point about CBT is this: it is long overdue for a feminist rewrite. Objective reasoning and emotional intuition are equally beneficial, and *both* belong in our arsenal of skills for better understanding ourselves and the world. At times we can challenge our assumptions, and at others we can honor our hunches and intuition, even in the face of incomplete information. Both offer a form of authentic knowing *and* a path out of anxiety.

A HOTLINE TO TRUE ANXIETY

If dreams are "the royal road to the unconscious," as Sigmund Freud once stated, then psychedelics are the divine hotline. A patient once described the experience she had on psilocybin—the psychoactive component of mushrooms belonging to the *Psilocybe* genus—as feeling "as if I was sitting in my therapist's office, delving deeper and deeper, only to look up and discover the psychoanalyst was *me*." I myself have experienced psychedelic ceremonies as more like having a therapy session with God. A number of my patients who have immense difficulty addressing their true anxiety or finding peace have ultimately found a path forward with psychedelics; they have discovered that taking these medicines not only allowed them to finally confront deep-seated fears but also to negotiate a truce with them, creating an enduring sense of calm.

You may remember my patient Ethan, in chapter 3, whose true anxiety stemmed from childhood trauma. He was prescribed Klonopin for panic attacks, and he had been trying to taper off for years. It had been brutal. I'd helped him try to slowly taper down, attempting to balance his significant anxiety with his desire to quit Klonopin. For years, it felt as if we were hitting a wall, and we both were growing increasingly discouraged.

The breakthrough, allowing Ethan to come off Klonopin completely and arrive at a new relationship to his anxiety, came when he "tripped on 'shrooms." Ethan had done psilocybin mushrooms a handful of times on his own in the past. And of his own volition, he'd come back to them when we were in the midst of this tortuous Klonopin taper. In our next session, he told me about a salient moment in his most recent psilocybin ceremony. He was sitting on a couch when he felt led to encounter an old trauma that had existed just out of his conscious awareness since childhood. Ethan recounted to me feeling drawn through a forest to a mass of cobwebs and vines. He struggled to push the vines aside; once he'd done so, he encountered a hidden door, behind which was a dark room containing a safe with a complicated lock. He continued to proceed through each of these safeguarding layers before reaching the tender center, which, when accessed, filled the space with flickering light, almost like a movie projector, showing him an ephemeral image of a vaguely familiar childhood experience. Ethan felt an urge to escape, but he resisted, reminding himself to surrender and allow the medicine to take him where he needed to go. He stayed with the image, the memory, and the complex array of emotions it brought up, from confusion and helplessness to rage and guilt. Ethan's body then went through a significant physical release, complete with shaking and rocking.

This brings to mind a hesitation that many people have about psychedelics—the fear of having a "bad trip." A challenging trip is not an inherently bad experience. It can be very, *very* difficult— I've been dropped to my knees on a number of occasions—but I generally trust that psychedelics take us where we need to go. I don't use the term "bad trip" because I don't think a difficult experience is necessarily a "bad" thing. I sometimes think of hard ceremonies as having the quality of a deep tissue massage. The catharsis can be beneficial. Just as in a massage, deep kneading in our most tender knots can be painful in the moment, but it leaves us transformed.

After some time shaking and releasing in this way, Ethan stood up from his couch. He described feeling as if his anxiety had been a heavy coat he'd been wearing for decades—and now he'd finally been able to slip it off and leave it behind. A couple of weeks later, Ethan stopped taking Klonopin as if it were no big deal, and for several months after that, he felt liberated from his anxiety.

Because therapy is rarely straightforward—and, sadly, is never a fairy tale—Ethan's anxiety reared up again, about three months after his psilocybin experience. I suspect that Ethan's brain is still recovering from Klonopin; that's how strong of a neurochemical hold I believe it had on him. At the same time, I also believe that the very growth that made it possible for Ethan to discontinue Klonopin has now landed him at a new and more difficult stage of psychological development. As if now that he has mastered one level of a video game, he has entered a more advanced phase. As Elisabeth Kübler-Ross and David Kessler put it, "The more you learn, the harder the lessons get."[15]

Research using functional magnetic resonance imaging to track activity in the brain when under the influence of psilocybin has shown that the drug decreases activity in two areas—the medial prefrontal cortex (mPFC) and the posterior cingulate cortex (PCC).[16] The mPFC, in particular, is known to be more active during depression, and the PCC is understood to play a role in consciousness and self-identity; heightened activity in this region correlates with excessive introspection, being stuck in one's own head and detached from the outside world. Psilocybin quiets this network down, allowing a person to "[beat] a new path in the brain," breaking them free from an "'overly reinforced trajectory.'"[17]

As such, Ethan's encounter with his trauma—his true anxiety—while on psilocybin afforded him an opportunity to push past a stage in his psychological development that he'd been previously unable to do.

There's reason to hope that psychedelic treatments could prove to be as major an innovation in psychiatry as the advent of SSRIs in the late 1980s. The field of psychiatry is currently in crisis. The SSRIs frequently prescribed for anxiety and depression are not as effective as we once believed, at least in less-severe cases.[18] They can carry a heavy burden of side effects and, as we've discussed, it can be difficult, even excruciating, to discontinue them. It's early days yet, but some very interesting recent research suggests potentially groundbreaking mental health treatments involving psilocybin,[19] ketamine,[20] MDMA,[21] LSD (lysergic acid diethylamide),[22] ibogaine,[23] and other psychedelics, which have been shown to help such debilitating disorders as anxiety,[24,25] post-traumatic stress disorder,[26] major depression,[27] eating disorders,[28] and opiate addiction.[29] Repeated trials have, indeed, shown impressive results. And unlike conventional medications, which create dependence over time, in many cases, psychedelics eliminate the need for themselves.

These substances, I believe, will increasingly come to be understood as revolutionary approaches for patients with intractable mental health issues. I'm glad we are headed in a direction that will make these treatments accessible to many more people, offering a chance at transformational healing. I also hope that as psychedelics become increasingly mainstream and overlap with the pharmaceutical industry and the medical setting, we can nurture a sense of honor and respect for them. As traditional cultures, such as the Urarina people of the Peruvian Amazon and the Pygmy tribes of central Africa, have known for several centuries, these medicines are sacred and, as such, should be approached with reverence and care.

Drugs: Illicit vs. FDA Approved

Under the right circumstances, I tend to think that cannabis is a safer sleep aid than Ambien, and psilocybin is a more effective antidepressant than Prozac. When we think about what substances are helpful and which are potentially harmful, it's important to consider the fact that the substances we're told are too dangerous for public consumption are *not* necessarily more harmful than medications that are approved by the FDA and on the market today.[30] There are political, economic, and historical drivers that determine whether a psychiatrist can write you a prescription for a drug or whether its use could get you thrown in jail. And this doesn't even begin to scratch the surface of the systemic racism at play in the criminalization of cannabis, with police and courts disproportionately enforcing such laws against African

Americans. The American Civil Liberties Union reports, for example, that Black Americans are up to eight times as likely as white people to be arrested for cannabis possession,[31] even though rates of cannabis use are similar in the two groups.[32] The real beneficiary of the criminalization of drugs, in my opinion, is not the public but the wine and spirits industry, the pharmaceutical industry, and the prison industrial complex. All of this is to say, don't automatically equate legal with safe and illegal with dangerous. I encourage you to evaluate each substance on its own and in the context of your unique health proclivities.

Considering the critical caveats that psychedelic treatments are not safe for everyone—they are relatively contraindicated for anyone with a personal or family history of bipolar disorder, schizophrenia, or other psychotic disorders (more research is needed); and I *do not* recommend taking these drugs outside a safe, facilitated setting or without support to integrate the experience—I do believe we have the neurobiological, psychological, and physiological reasons to be hopeful as well as to fund more research.

Psychedelic medicines improve the functioning of our brain chemistry in a variety of ways. They enhance serotonergic signaling at what are called the 5-HT_{2A} receptors in the brain in a way that seems to be lasting and effective, without being numbing or creating a withdrawal state.[33,34] This is what researchers believe accounts, at least in part, for the enduring antidepressant and antianxiety effect of a single psilocybin ceremony. Certain psychedelics also increase the secretion of a very important signaling molecule called BDNF, or brain-derived neurotrophic factor, which promotes neurogenesis and neuroplasticity.[35,36] Translation: BDNF helps the brain grow, change, and adapt;

so, if you're stuck, as Ethan was, BDNF helps you get unstuck. This discovery has exciting therapeutic implications for treating entrenched psychological patterns such as PTSD, ruminative depression, and addiction. Certain psychedelics are also anti-inflammatory[37]—which is useful because, as you understand well by now, inflammation is a common contributor to anxiety and depression.

Another line of research has explored the effects of these medicines on the default-mode network (DMN)—composed, in part, of the aforementioned medial prefrontal cortex (mPFC) and the posterior cingulate cortex (PCC), the parts of our brain responsible for our sense of ourselves as separate from others—suggesting treatment possibilities for people experiencing alienation, loneliness, and trauma as well as the anxiety that comes along with these. Viewing ourselves as individuals navigating our own challenges has served the human race to a certain extent evolutionarily, allowing us to learn from our mistakes, anticipate potential negative outcomes, and fight for our own survival. Yet a temporarily quiet DMN—which psychedelics help us achieve[38]—grants us a reprieve from future tripping and dwelling on the past. And perhaps we benefit collectively when more of us spend time with a dialed down DMN, allowing us to explore a feeling of interconnectedness with our fellow beings and our planet, while rethinking that narrow definition of ourselves as separate.

Finally, psychedelic experiences can more directly help people discharge stress, allowing them to complete the stress cycle, through shaking and sounding (e.g., chanting)—actions that are common in the psychedelic experience but typically inaccessible in our relatively more inhibited daily lives. In less clinical terms—and, frankly, from my own experience taking ayahuasca in Brazil (where it is legal) and psilocybin in a formal setting with a facilitator—for a few hours during the psychedelic ceremony, you get a glimpse of another reality, which can take the pressure off this one. It's humbling and freeing to see beyond logic that there is more to life than our material existence—and that perhaps it's not all up to us. You might begin to trust that something greater than what you could possibly comprehend is unfolding. "Psychedelics are much more than tools for healing trauma," as my friend and colleague Will Siu, MD, PhD, wisely remarked. "Psychedelics are helping spirituality become palatable to a starving Western world."[39]

Perhaps most critically, for some, psychedelics can decrease a fear of death—which sits at the very center of true anxiety. Several randomized controlled trials, with longitudinal follow-ups, have shown that psilocybin not only inspires spiritual revelations, reducing anxiety and depression,[40] but also allows people to overcome end-of-life anxiety. In 2011, for example, researchers at UCLA conducted a psilocybin trial led by psychiatrist Charles Grob, with twelve terminal cancer patients experiencing anxiety, depression, and existential dread. Researchers found a significant reduction in anxiety and in fear of death among these patients after three months—and an *improvement* in mood for up to six months afterward.[41] This study was corroborated by a large randomized, double-blind study by researchers at Johns Hopkins University in 2016.[42]

Still there are those who assert that these awe-inspiring experiences are simply based on neurobiological effects of psychedelics—and that there is always a rational explanation. I have a group of friends, for instance, who are atheist engineers in California. In their spare time, they are "psychonauts," experimenting with psychedelics. When we catch up from one coast to the other, they tell me stories of their ceremonies. Although, subjectively, they experience an extraordinary sense of something beyond themselves—some communion with the divine, a transcendent feeling of awe, or a cracked-open moment of compassion for someone they have struggled to forgive—they come back from these encounters trying to understand the neurochemical correlate to explain it. I understand their resistance to believing in something divine, and ultimately as we learn more about the therapeutic potential of psychedelic substances, I expect there will be more pressure to determine the neurochemical basis for such experiences.

This brings to mind the so-called mystical experience hypothesis, which posits that when someone takes a psychedelic in a proper set and setting, they will reliably have a peak spiritual experience, typically feeling a sense of unity or a recognition of oneness with the world.[43] And the degree to which someone has a mystical experience predicts positive therapeutic effects—that is, the mystical nature of the experience is proportional to the lasting benefits of the psychedelic (such as a decrease in depression).[44,45] In other words, the more mystical the experience, the more effective the medicine. *National Geographic* host Jason Silva describes this phenomenon as "inverse PTSD"—an experience of such radiance, astonishment, and grace that it can transform a person's character structure in a similar way to that of trauma but building toward openness and love rather than fear and mistrust. The mystical experience hypothesis suggests that *the journey itself* is what offers much of the benefit. This is especially interesting when we consider the implications for the "pharmadelics" that are sure to be coming down the pipeline. When the pharmaceutical industry inevitably tries to isolate the active ingredients and provide us with the benefits of a psychedelic ceremony without the messiness of a "trip," it may well be inherently less effective. There's a reason we call it a "trip"— psychedelic medicines guide us through a journey, and we learn and grow along the way. In the words of my friend, psychedelic researcher and psychology professor at Yale, Alexander Belser, PhD, the messy part of the trip is "a feature, not a bug."[46] I tend to believe this is true not only with psychedelics but also in life.

Whether through meditation or psychedelics, the most important first step toward tapping into our inner wisdom is to *listen* for it. Once this deepest part of yourself makes itself known, trust its message, as this is your own unique and essential guidance, advice that only *you* can offer—and take.

Chapter 13

This Is Why You Stopped Singing

Burnout exists because we've made
rest a reward rather than a right.

—*Juliet C. Obodo*

Most of us know the story of the canary in the coal mine. But, just in case, here's a recap: Throughout much of the twentieth century, coal miners used to bring canaries in cages down into the mines with them as a way of detecting carbon monoxide, an odorless gas that can build to deadly levels in that environment. Canaries are more vulnerable to airborne poisons than humans,[1] so when the birds stopped singing, the coal miners knew they needed to get out. This has since become a well-worn metaphor for something—or someone—whose sensitivity to adverse conditions offers a warning cry of coming danger.

If you struggle with anxiety, there's a good chance that *you* are the canary in our coal mine. That is, you're sensitive enough to have detected the toxic influences of our modern world and, perhaps, you too have stopped singing. There are many different terms for this type of person: empath, intuitive, highly sensitive person (HSP), artist, healer. It means you may have a bigger antenna than the average person, so you pick up more of the background noise. This can be a liability, because the world can be pretty loud these days, but it's also a gift. So, if you're dwelling on the negative aspects of being sensitive—that you're particularly emotional, or you can't handle crowds, meet-and-greets, or gluten—remember that you also possess the positive attributes of this trait. You're likely more attuned to others' needs, and better able to listen on multiple valences at the same time, hearing not just someone's words but also picking up on the way they're holding their body or betraying their hidden moods. And of course, you have your ear to the ground in terms of the larger needs in the world. Sensitivity is a calling and it should be prized—and handled with care.

The Sensitive Person's Daily Checklist

We all need to brush our teeth and drink water. But if you're sensitive, you may need some extra maintenance to care for your inherently perceptive nervous system. Here are a few other practices to consider adding to your daily routine.

- Go to bed early and, when possible, wake up without an alarm.
- Give yourself quiet and solitude when you need it.

- Simplify your life wherever possible—don't overcrowd your calendar; say no when you need rest.
- Ground yourself in nature—spend at least ten minutes a day with bare feet touching the ground.[2]
- Clearing practices—if you've taken on a lot of energy from others, do a few minutes of an energy-clearing practice, such as shaking to shamanic drum music.

Humans are variable by design. As evidenced by the primate study discussed in chapter 3, the sensitive members of our tribe are necessary for our survival. They are on the front lines, sending warnings back to the rest of the pack. In the past, they saved us from storms and stampedes; today, they are making us aware of the ways in which the world is perilously out of balance. "We, the highly strung, are the advance party who flag to the troops that consumerism is hurting our hearts," as Sarah Wilson writes in *First, We Make the Beast Beautiful*.[3] On a more personal level, sensitive people are often tapped into the quiet misgivings of others, such as when a person in the room isn't being heard, or when someone feels upset; these are the people who shift the energy of the room so that it feels more gentle and equitable for all. In addition to our intuitives, however, we also need those who have a bit more evenness and durability—our surgeons and pilots, for instance—to serve different functions. We're all built to serve a role in society, whether we're unflappable under pressure or unable to make it through the news without crying. The truth is, though, we are *all* capable of losing our song, at least for a while, especially in our current world. If you could drop a litmus strip into the stew of modern Western culture, it would reveal that the tone of life these days *is* anxiety. Anxiety is the verb, the vibe, the texture, the pH of our age.

THE BANALITY OF FEAR

*It did what all ads are supposed to do:
create an anxiety relievable by purchase.*

—David Foster Wallace, Infinite Jest

In addition to our current heavy-duty societal causes of anxiety—such as systemic racism, climate change, and the trauma of sexual assault and harassment—there are also so many intimate reasons that we live in a state of worry. We are worried about our jobs, our marriages, our families, our lack of relationship or family, our health, our finances. And layered over these is the fear that we are constantly bathed in, by something as seemingly benign as advertising.

Forget sex—large corporations have figured out that *fear* and anxiety sell. Since contentment and self-acceptance don't exactly inspire rampant consumerism, our insecurities are carefully curated and reflected back to us instead, instilling uncertainty and driving us to consume. As we scroll through our targeted advertisements on social media, we repeatedly encounter the message that we're not enough, we're in danger, something urgently needs fixing. As a result, we go through our lives bathed in fear—not necessarily for a deep, sinister reason, but simply because clever marketers are trying to make a buck. Our anxiety epidemic has been amplified by something as banal as marketing strategy. If you find that you're losing your song, start to notice anytime you're being sold something. This awareness can act like a force field, preventing you from getting swept up in the anxiety-provoking messaging that was merely trying to scare you into buying something you don't need.

ALWAYS ON THE CLOCK

A reckoning with burnout is so often a reckoning
with the fact that the things you fill your day with—
the things you fill your life with—feel unrecognizable
from the sort of life you want to live, and the sort of
meaning you want to make of it. That's why the burnout
condition is more than just addiction to work. It's an
alienation from the self, and from desire.

—*Anne Helen Petersen,* Can't Even

For years we have used the term "workaholic" to refer to someone who seems to be addicted to their job. But I think of workaholism as a state in which people use the office as an escape, a way to distract from troublesome emotions, such as avoiding facing a disintegrating marriage or the difficulties of raising children, all while believing we're building *toward* something—namely, wealth and status.

Today, however, our compulsive work habits seem to have a different root. The never-ending workflow is not workaholism but instead workism—"the belief that work is not only necessary to economic production, but also the centerpiece of one's identity and life's purpose," explains Derek Thompson in his *Atlantic* article, "Workism Is Making Americans Miserable."[4] I get the sense that many of my younger patients don't want to escape their feelings through work with the same urgency that people once did; on the contrary, they are *invested* in work giving their lives substance. Research has indeed shown that millennials have more of a drive toward finding meaning and aim in their work.[5,6,7] Meanwhile, start-ups and tech giants alike have figured out that if they hide behind the fig leaf that they are "making the world a better place," they can use this battle cry to convince young employees to accept lower pay *and* work non-stop, because if we don't act as team players, we're not *committed to the mission*. "There is something slyly dystopian about an economic system that has convinced the most indebted generation in American history to put purpose over paycheck," Derek Thompson writes. "Indeed, if you were designing a *Black Mirror* labor force that encouraged overwork without higher wages, what might you do? Perhaps you'd persuade educated young people that income comes second; that no job is just a job; and that the only real reward from work is the ineffable glow of purpose. It is a diabolical game that creates a prize so tantalizing yet rare that almost nobody wins, but everybody feels obligated to play forever."[8]

And for most of us these days, we would like nothing more than to take a month off to rest and spend time with the people we love. Yet we never get that month or week or even day off, because . . . We. Are. Always. Working. We scarf down our lunch while staring at Slack. At night, when we're "relaxing" and watching TV, we're still answering work emails from the phone, and we're sharing the couch with a spreadsheet. We bring our laptops on vacation, and we bring our in-box into the bathroom. Then, from our beds, we ponder work predicaments while struggling to fall asleep.

There once was a time, believe it or not, when there were contextual cues—and a consensus—that the workday was over. First, a steady trickle of people packing up, punching out, and leaving, and eventually the fluorescent ceiling lights shutting off . . . This series of events offered a fail-safe communication: it's time to go home. Now, our work lives are always in progress, without a fixed endpoint. Today, for many of us, our office is a laptop on the dining room table, and we work continuously, morning till night. We already had a burnout epidemic on the rise in this country[9]—and then we entered the COVID-19 pandemic, further blurring the boundaries between work and home but also adding a heavy dose of uncertainty, grief, and collective trauma to the mix. Parents at least have their workdays punctuated by the needs of family and bedtime routines; although, as far as I have gathered from my patients (and let's face it, my own life), many of us are inevitably signing back on after our children are in bed in order to meet the demands of our jobs.

This cultural shift toward workism has made us very productive—and also very *anxious*. Our exhaustion, as Brené Brown has pointed out, is now viewed "as a status symbol" and our "productivity as a metric for self-worth."[10] Our constant availability through technology has robbed us of a sense of accomplishment, and, instead, we are in a ceaseless race to get to the end of an infinite to-do list. Even for people with relative abundance, there is no more *enough*. Anxiety is in many ways an exaggeration of our survival instinct—forage for food, prepare the nest, and be on the lookout for predators and disasters. In the modern work world, this drive toward preparation and survival can go on endlessly. And the resulting feeling—that we could always be doing more—is anxiety catnip.

As such, even our relaxation has taken on a sense of striving. That is, we meditate for better focus; we go to bed early so we can be clearheaded for our meeting in the morning. In Chinese medicine, the Taoist concept of yin and yang expresses the idea that the world is composed of two opposite but mutually interconnected forces. Yin is darkness; yang is light. Yin is feminine; yang is masculine. Yin is rest; yang is activity. The idea being that everything in the world is in natural and dynamic equilibrium. But today we have tilted the scales so heavily toward work and productivity that we have come to only value yin when it is *in service of* yang. We now see leisure as a function of increasing output.

And therein lies the rise of the wellness industry—or the "productive relaxation" industry.[11] Given our need to accomplish every minute, even when we're at rest, the widespread appeal of hyperdriven self-care starts to come into focus: meditate in an infrared sauna; practice gratitude while doing a headstand on a paddleboard. Currently, our yin is really just more yang, and our leisure itself has become exhausting. And yet these two sides of ourselves are meant to be in equal balance, meaning we should value leisure for the sake of leisure—as much as we value industriousness.

Ironically, the fact that we've debased leisure has actually caused us to have *less* dynamic work lives. We are distracted and prone to procrastination at work precisely because we never really rest. If when we are supposedly relaxing, we are also checking work email or squeezing in a regimen of intensive self-care, how can we expect to gain the restoration we need to later engage in significant work? We're only ever half off when we're supposed to be off, so we're only ever half on when we're supposed to be on, creating a vicious cycle—we lack that sense of satisfaction for a job well done during the day, and we feel even more pressure to produce when at rest.

It's important to remember: your manager, your company, and even your lifelong unconscious conditioning will not encourage you to rest. The onus is on us to consciously and proactively designate a time for leisure—and then protect it fiercely. First, begin the day by setting the tone *for yourself.* What I mean by this is don't pick up your phone first thing in the morning.

Don't let your phone—with all its demanding notifications—tell you how to feel. Wake up and be with yourself and your own mind long enough to establish an intention and an atmosphere for the day. Then, go outside—even if it's for just two minutes, standing on the sidewalk in pajamas—and catch a dose of actual sunlight. This starts the clock on your circadian rhythm, kicking off a symphony of hormones helping your body know that it's daytime—time to feel awake, alert, and engaged. It also begins the countdown to help you feel sleepy at night. A few minutes outside also provides some separation between work and life, and it builds just a bit of spaciousness into your day.

When it's time to start your workday, jot down a schedule with realistic goals and designated breaks for meals and simply resting your brain. Scrutinize the need for every meeting and keep an eye toward protecting large blocks of time for dropping into a flow state (that state of feeling immersed in a task with energized focus) and accomplishing substantial work. As leadership and business strategist Greg McKeown presents in his book *Essentialism: The Disciplined Pursuit of Less*, the question is, do we want to "make a millimeter of progress in a million directions" or a mile of progress in one direction?[12] Perhaps most importantly, decide at the beginning of the day what time you're going to stop working. If you then arrive at the designated time and you're not yet finished, you have my blessing to snooze the alarm on this once. Put out the last fire, ship out the last item, then shut it down. Finally, it's also critical to have a ritual that signals to the brain that you're done with work and it's time to relax. Take a walk; dance in your living room; sip tulsi tea while watching the sunset. It doesn't have to be elaborate, but it should be intentional (i.e., not just zoning out on social media).

Our culture is addicted to being busy. We exist in a perpetual state of feeling as if our to-do list is infinite and we have no time. This scarcity mindset is chipping away at our ability to adequately rest *and* work. And the truth is, after years of living this way, we are *exhausted*. Reclaim time in your life—five minutes, then ten, maybe a whole afternoon—to take a walk, to do nothing. This will start giving your brain the signal of abundance; that you have enough, that you *are* enough.

Anxiety of Wellness

Where we think we need more self-discipline, we usually need more self-love.

—Tara Mohr

Ironically, the pressure from the wellness industry to be—and buy—our "best selves" is making us anxious. "Self-care" practices promise to make us feel good and virtuous, but they come with a hidden stinger: they are yet another item on our to-do lists, another list of acquisitions to drain our bank accounts and clutter our lives, and another message that we are not enough as we are. By asking us to do and buy too much, we are set up for failure and overwhelm, and the attendant guilt and anxiety. And, most importantly, the implicit conceit of "wellness" is that we are broken and need fixing, when in fact, we are inherently *whole*. Real self-care is self-love, community, nature, and rest. The pressure to do all these other elaborate rituals can actually undermine your well-being. If the self-care industrial complex is only adding to your worries, unsubscribe.

THE ANXIETY OF ACHIEVING SOMEONE ELSE'S GOALS

My patient Hanh is the daughter of immigrants who came to the United States as refugees fleeing a Communist takeover. Her parents arrived with next to nothing, and they hustled to survive, driven by a desire to create a better life for their daughter. Hanh, thirty-six-years old, works as back-office support at a bank. She makes good money, and her parents are pleased with her success.

But Hanh is miserable and anxious. She doesn't love the work, and she feels trapped under the weight of her parents' expectations. She speaks to me often of quitting her job and pursuing a career in early childhood education. She also knows, however, that her parents would see this as an outrageous thing to do; they would view it as financially unwise and a betrayal of everything they've fought so hard to provide for her.

When Hanh's parents emigrated, they were operating under a system of *legitimate* scarcity. Under those circumstances, focusing nearly all their time and energy on earning money was the right prioritization. They wanted to secure their future as well as their daughter's future, and they succeeded. Hanh was given a foundation from which she was able to get a good education, earn a good living, and put away some savings. As a result, she does not exist in the same system of scarcity that her parents did, as I often remind her in our sessions; Hanh has relative abundance. Thus, her priorities *can* and *should* be different. If she is focusing solely on making as much money as possible to secure a future at the expense of her happiness and well-being, then what was it all for? I've seen this time and again with my patients: they are striving for a goal that isn't even of their own making.

Hanh still works at the bank. But she has heard the clarion note of her true anxiety—she is beginning to recognize that she has been blindly working toward someone else's goal. She is now making careful plans to leave. She is terrified of disappointing her family or seeming ungrateful. Our conversations have focused on honoring what her parents went through while still finding the conviction to boldly step out from under their values. I remind Hanh that the goal is to design your life with intention—bringing deep, conscious reflection to the choices you make about work, salary, passion, prestige, responsibility, purchases, and rest. Instead of defaulting to what the world is asking of you, consider everything that is hanging in the balance and make choices *for yourself.*

DROP PERFECTIONISM

Perfectionism is internalized oppression.

—*Gloria Steinem*

Perfectionism is a coping strategy—it's a way we try to claim our seat at the table. But it turns out that it only serves to paralyze us, and we already inherently deserve a seat at the table, even if we're imperfect. *Especially* because we're imperfect.

First, consider why perfectionism is a priority for you. Is it a product of outside forces or early life experiences suggesting that you must always do and be better? In your childhood home, was attention wrapped up with achievement? If you felt that you earned your parents' love through impressing them, then you very well may hold yourself to exacting standards in adulthood.

Step back and ask yourself, Do you believe you need to achieve or please others to be worthy of love?

Second, it's important to realize that everything is a trade-off. Many of us are trying to be perfect at all things at all times—work, health, activism, and showing up for kids, partners, friends, and aging parents. Acknowledge that anytime you put more of your energy toward one thing, something else is inevitably compromised. It's impossible to get this perfectly right, so we can drop that as a goal.

For many anxious folks who struggle with perfectionism, the belief is that if you're not bending yourself into a pretzel and burned out, then you're not doing your best. Here's a new definition of doing your best: do your *reasonable* best. What is a reasonable effort that you can feel proud of, that you can achieve while still staying rested, calm, and balanced? That's your reasonable best, and having this as your goal will allow you to crawl out from under the tyranny of perfectionistic standards and go through life with considerably less anxiety.

Sunday Scaries

For many with anxiety, the Sunday scaries—"the flood of anxiety that many of us feel as the weekend is winding down and the workweek approaches"—according to journalist Derek Thompson, is when "a psychological tug-of-war" takes place between "productivity mind" and "leisure mind."[13] An important consideration is to recognize that sometimes the Sunday scaries are a product of false anxiety—precipitated by later bedtimes, a Saturday night bottle of wine, and hitting the French press a little harder on Sunday morning. If you've played it fast and loose with diet, intoxicants, sleep schedule, and scrolling over the weekend, your body will be grappling with a cluster of physiological stress responses just as you're facing the prospect of going back to work. But sometimes that Sunday evening dread has a true anxiety quality. Once you take steps to rebalance your physical health and reduce false anxiety, you'll be in a good position to discern what your inner compass may be pointing you toward. Sometimes Sunday scaries are a product of your soul rebelling against a job that is out of alignment with your purpose or values. Sometimes we dread Monday morning not because we were reckless over the weekend but because our work feels intangible, disconnected from a vital contribution. I've seen enough patients who have sifted through their various false anxieties on Sunday evenings, only to find an insight buried beneath the stress response. I've come to realize that we all need to be listening carefully to the truth behind our Sunday scaries.

Process over Outcome

We are an outcome-focused society—centering our attention on grades and income, social media likes and followers, and generally how the world receives us. But what if we shifted our focus away from the outcome and instead to the process? That's the only part we're responsible for—and the only part we actually control. We have next to no dominion over how we'll be perceived in this life. That is subject to everybody else's internal biases and to chance and luck as well as all sorts of misunderstandings along the way. Not to mention, if we focus only on how we are seen, we end up shape-shifting to fulfill the expectations of others. This inevitably becomes a game of people-pleasing whack-a-mole. As soon as you please one person, someone else will feel let down. When you shift to please that person, you'll disappoint someone else. And in trying to win this game, you will inevitably betray yourself and feel a nagging sense of anxiety that something isn't right. So, release yourself from the stress of white-knuckling everyone else's impressions and just show up, do your *reasonable* best, and remain unattached to the outcome.

If you are one of the sensitive members of our human race, one of the canaries in our collective coal mine, it's likely that some insidious aspects of modern life have been slowly poisoning you, dragging on your spirit, and instilling a sense that you will never do, have, or be enough. It's imperative that you learn to turn the volume up on the voice of your true anxiety, letting it overcome the noise pollution of our culture, and use it to guide you toward fresher air. Ultimately, you will carve a better path for us all.

Chapter 14

Connection Is Calming

The seventy-five years and twenty million
dollars expended on the Grant Study
points . . . to a straightforward five-word
conclusion: "Happiness is love. Full stop."

—*George Vaillant*

I could talk about the ways sleep deprivation and inflammation
are impacting our anxiety until the cows come home; I truly
believe these factors make a significant difference in the way we
feel. But every additional year that I do this work, I learn that,
when it comes to our mental well-being, few things matter more
than the relationships in our lives. If you have an opportunity to
sit around a dinner table talking to people you love, communing
and laughing until 2:00 a.m.—while eating bread and pasta and
drinking wine—that may be a better choice for your health than
declining the invitation, eating all the right foods, and crawling

into bed by ten. At the end of the day, our well-being rests on our connections with other people above everything else.

But you don't need me to tell you this. Every religious text, mushroom trip, and sophomoric poem will offer the same revelation: the answer to our growing discontent is love. And yet, despite the fact that the answer is easy, the real test is how we find it in our lives—how we show up, treat each other, offer ourselves in service to the world, and make our unique contribution in life.

We are hardwired for community. Allowing for some variations among us on the spectrum from introverted to gregarious, the evolutionary fact is that as a species, humans can't opt out of social connection, at least not without cognitive decline,[1] decreased longevity,[2] and anxiety.[3,4] As we've discussed, for millennia, our survival hinged on being a part of a tribe, and for this reason, community is a genetic directive. When we are without it, it's adaptive for us to feel uneasy until we're once again surrounded by our people. And we are not the only mammals wired in this way. We can still see the tribal instincts of dogs and wolves as they roam in packs. And even rats have been found to have a predilection for social connection: in a 2018 study led by Marco Venniro, PhD, of the National Institute on Drug Abuse, "addicted" rats were offered the choice between social interactions with other rats or heroin and methamphetamines—and they consistently picked community over drugs.[5] The reverse has been demonstrated as well—that is, rats put in isolation began to take *more* drugs.[6] Of course, as the researchers behind this study pointed out, human beings have more complex social needs than rodents, but these findings still provide valuable insight not only into the structure of recovery programs but also, more broadly, into the innate need for company and connection. And yet community is easier said than created.

FINDING YOUR PEOPLE

When we feel misaligned with the people who surround us or have a paucity of good people in our lives—not enough generous listeners worthy of our friendship—we become increasingly lonely and anxious. I encourage my patients to open their eyes to the other possibilities, to understand that there are *so many* communities in this wide world, but they might have to go exploring beyond their coworkers or college friends. A helpful approach can be to seek out new friends in places where people are trying to be better. This might mean making friends at a community meditation or a Twelve-Step meeting.

The flip side of forging new connections, however, is using your true anxiety to help you discern your limits in your existing relationships. I have patients who remain loyal to historical friendships, ignoring their true anxiety that has been tapping them on the shoulder saying, *This relationship is no longer the right use of your time and energy.*

I also have patients with a tendency to overcorrect in the other direction. That is, when a worthwhile relationship goes through a challenging patch, they can be too quick to call it toxic and set an aggressive boundary, when really they should be using boundaries to *salvage* relationships that have taken a wrong turn. The reality is that setting healthy limits shouldn't be used as a mechanism for rejecting others or walling ourselves off from interpersonal work; it should be used as a way to *encourage* a connection that honors our needs. Instead of setting a boundary in a punitive way, as in, *I'm denying access to me as a punishment for your bad behavior*, we can set boundaries from the perspective of *I really want our relationship to work, but there's something in how we're showing up for each other right now that is damaging our relationship*. Then place the boundary in an effort to protect the relationship and set it up for success later. In this second version, we're *rooting* for the relationship to work. In this sense, our boundaries can promote connection *or* separation; it's up to us to decide which one we want to fight for in our lives.

A Note about People Doing What They Should

We all want people to do what they *should*. This person *should* apologize to me; that person *should* have sent me a thank-you card for my gift; that guy *should* know better than to support that political candidate.

Breaking news: humans, throughout history, have not done what they *should*. And here's a wild prediction for the future: they're not about to start. All this waiting for people to start doing what they *should* is actually a form of resisting reality. It not only causes a great deal of suffering but also blocks us from enjoying our relationships. Here's a better strategy: accept how things are right now, and work from there, recognizing that people can't read our minds, and everybody thinks they're the good guy. "I am a lover of what is," writes author and founder of The Work, Byron Katie, "not because I'm a spiritual person, but because it hurts when I argue with reality."[7] By ending your long-standing debate with reality, you will free yourself up to accept people as they are and start liking them again.

CONNECTION OVER PERFECTION

The quality of our relationships determines
the quality of our lives.

—*Esther Perel*

If we really want connection in our lives, we need to allow for the messiness that comes with community. People are challenging, they say the wrong things and can be careless and insensitive, they leave the cabinet doors open, and they move your stuff. But meaningful connection with people we love is also, at the end of the day, not only a biological imperative but the bottom line for a fulfilling life.

My patient Noor is a thirty-eight-year-old mother of two. She works from home in digital marketing, and her husband frequently travels for work. Even though she is constantly fending off toddlers from invading her personal space while she struggles to work, she describes herself as "desperately lonely." She's starved for adult interaction and yearns for authentic connection, not just with her colleagues on Zoom. She tells me, "I want to connect with my girlfriends the way I used to before I had kids."

But when I suggest that Noor go out or invite people over, she pushes back, saying that coordinating and paying for babysitters is almost more trouble than it's worth, and the house is a mess. "I have zero time to cook," she counters, "and I always look like I just got hit by a truck." I recommended that Noor lower her bar for hosting in order to bring community back into her life. If we insist on cleaning up and cooking a multicourse meal every time we have people over, we'll likely see friends twice a year. Yet, for the nervous system to feel calm and safe, we need to feel held by community on a regular basis. It's a beautiful thing to pour love, time, and effort into making a meal for someone if you have the time and inclination. But Noor doesn't, and the expectations she was putting on herself were standing in the way of getting her needs met. So, I suggested that she lower her standards and just tell her friends, *The house is a mess, we'll be ordering takeout, but come on over anyway!* With this Noor started having her girlfriends over a few times a month. She wore her sweatpants, and they all sat among the mountains of Legos in her living room, eating tacos and laughing till they cried. Noor's anxiety has improved considerably now that she's getting her fundamental need for connection met on a regular basis.

EARLY LIFE AS A BLUEPRINT FOR COMMUNITY

Adverse childhood experiences have a lasting impact on the way we show up in relationships as adults. If our parents modeled emotional immaturity or bad communication or if we experienced trauma early in our lives, we are more likely to suffer in our own relationships. I've seen this pattern play out often with my patients who grew up witnessing unhealthy relationship dynamics—they become stuck in bad relationships, or they don't feel compelled by relationships with healthy boundaries. Sometimes they grow so frustrated by repeating the same painful patterns that they simply stop trying to connect with others altogether. If this sounds familiar, you may want to get reacquainted with your child self. We can travel back in time, therapeutically speaking, and tend to our younger selves to create better conditions for satisfying social connections in adulthood.

While we can't choose our parents, we *can* "reparent" ourselves. When suffering from the long-term effects of detrimental or neglectful parenting, it's important to look generously on how we once navigated our lives as young people. We were doing what we could, some of us scrambling to survive, with a child's mind and without the capacity to make sense of what we were going through. Behaviors that were understandable and adaptive *then*—such as putting others' needs before our own in order to "earn" love or closing ourselves off from depending on people out of fear of being hurt—may be maladaptive *now*. Thankfully, it is possible as adults to tell our younger selves what we needed to hear back then—that we are lovable and that we didn't do anything to deserve bad treatment.

I work with a forty-six-year-old man named Hector who grew up with a physically abusive older brother. This went beyond the normal brotherly wrestling and sparring, where my patient would often feel genuinely in danger. When he would report to his parents, time and again, that his brother had physically assaulted him, they countered with invalidating remarks like, "Well, you must have done something to provoke him." They were clearly out of their depth in managing the older brother's behavioral issues. Ultimately, Hector's brother was diagnosed with conduct disorder (a serious behavioral and emotional disorder in young people), but not until Hector had already internalized much of the violence and blame. In his therapy with me as well as with the support of a long-term relationship with a wonderful partner, Hector has successfully reached his boyhood self, relieving him of the feelings of accusation and guilt, realizing he was justified in his fear, and allowing him to understand he is a good person worthy of a happy life.

And yet still today, this early childhood blueprint can make Hector feel unsafe. This comes up most often as health-related anxieties. When Hector feels his body is in disrepair, he sees a round of doctors, seeking endless reassurances. When he describes these experiences to me, I can hear the distant emotional echo from his childhood, beseeching: *I need protection. Can you help me?* Because in childhood, this question was met by victim-blaming from his parents that overlooked the very real danger Hector was in, he was taught that he couldn't trust the grown-ups. Today, doctors are stand-ins for the untrustworthy adults in Hector's childhood, and he struggles to accept their reassurances as the final word.

Fortunately, Hector has found therapeutic modalities that help reprogram those patterns laid down by trauma. Remember the rabbit and the wolf example in our discussion of the stress cycle in chapter 11? This cycle also has profound implications for the treatment of trauma. When a trauma occurs, our body goes through a significant fight, flight, or freeze response.[8] If this stress cycle is not brought to completion, adrenaline can continue to flood the body, leaving us with an intense energy that needs to be discharged, and a chronic state of hyperarousal, which can continue as long as the trauma is held in the body. It's as though the unmetabolized trauma is a key stuck in the ignition of your fight-or-flight response, keeping the engine idling. In that state, your unconscious is hypervigilant, forever surveying the landscape for threat. Thoughts and feelings are filtered through the fear-tinted lens of trauma. Interactions and sensations are perceived as more menacing than they actually are—similar to Hector fearing for his health and not being able to trust medical opinions. This is something like the true-anxiety version of the stress cycle and, as such, it's deeper and more difficult to complete.

Trauma is held in the body, in the connective tissues and in the nerves and fibers of the nervous system, and therefore talk therapy is limited in its usefulness. In fact, hashing and rehashing a trauma verbally, even with a caring therapist, has the potential to be *retraumatizing*. The best treatments for relinquishing long-held trauma—and finally releasing its ongoing stress cycle—are trauma-specific therapies. These include eye movement desensitization and reprocessing (EMDR), in which patients close their eyes and move them rhythmically as a therapist talks them through their traumatic events (the mechanisms at play here are still not entirely understood, but clinical trials conducted over the last several decades have proven it to be successful);[9,10] Dynamic Neural Retraining System (DNRS), which aims at reprogramming the limbic system; and Somatic Experiencing therapy, which uses mind-body exercises to release trauma. These therapies take into account the physical need for moving the stuck energy of the stress cycle and accessing trauma at the level of the limbic system and the body, making them much better suited to trauma recovery than traditional talk therapy is. Trauma-focused therapy can also be especially helpful in allowing the brain to register the message, *That was then, this is now, and you are safe.*

ATTACHMENT AND RELATIONSHIPS

Another way that our childhood impacts our adult relationships is in shaping our attachment style. Attachment theory, based on the work of developmental psychologist Mary Ainsworth (1913–1999), posits that caretaker responses in early childhood create our attachment patterns—such as secure, avoidant, anxious, or disorganized—which go on to influence our adult relationships. When children grow up with caretakers who are sensitive and responsive, they can develop secure attachment, meaning they can experience a primary relationship as a safe and reliable home base. What I commonly see in my practice are people whose caregivers failed to properly attune to their emotions and needs in early childhood. Sometimes this occurs because a parent or caregiver had mental illness or struggled with substance use; often caregivers have their own trauma they were still working through; sometimes life just throws a household a curveball and parents become too preoccupied with the stressors, unable to tune in to their children's emotional needs. All these can create an anxious attachment style, which can manifest in adulthood as difficulty with trust and fear of abandonment.

I have also seen anxious attachment in adulthood become a self-fulfilling prophecy. That is, it can cause people to perceive insult and offense where they do not exist, ultimately driving people away. I have a patient named Zahara, for example, who grew up with a mother with mental illness who was too depressed to properly respond to Zahara's cues. She was also hospitalized for a month when Zahara was three years old, which led Zahara to believe, in that magical-thinking way of the toddler mind, that she did something to drive her mother away. Now Zahara goes through adulthood thinking that her emotions are too much for others to handle. So, in an attempt to prevent abandonment, Zahara finds ways to obligate people into spending time with her—by guilting them into it or making them feel sorry for her. While this gets some people to reluctantly show up, it has also generated resentment and driven people away in the long term, *creating* the very abandonment she dreads—and turning the screw once more on her original insecurity.

If this sounds familiar, the solution lies in acknowledging that your early childhood relationships are shaping the way you approach adult relationships. You may have sought out unreliable partners in an unconscious attraction to what feels familiar. At the same time, you may have passed by a partner with a secure attachment style because it felt like there was "no chemistry." Going forward, when someone demonstrates themselves to be a securely attached, reliable partner, do some careful experimentation around trusting them, as vulnerable as that may seem. This will ultimately create a more abiding sense of safety and calm in your relationships and life. And if you identify manipulative practices you might be employing in an unconscious bid to prevent abandonment, see if you can try to relinquish the need to control others and allow them more freedom. Though you might fear that this will cause people to leave, the truth is that it will more often allow them to stay. Throughout the work of learning how to show up in relationships in a new way, continuously remind yourself that, though you have these wounds, they are undeserved, and you have always been—and always will be—inherently worthy of love.

DISCONNECTED FROM SOURCE

Though the COVID-19 pandemic brought the term "social distancing" to the forefront of our cultural lexicon, we've been sliding into isolation over several decades—since the advent of the smartphone, the internet, the Walkman. And along the way, we've been progressively weaving isolation into our lives not only through technology—pulling us deeper into a digital community and away from the lifeblood of real, live community—but also through a broader disconnection from our spiritual selves, or what I call source. I use the word *source* to convey that which stirs our sense of wonder, such as nature and creative expression. This increasing detachment has a psychic cost in that it carries a rising swell of anxiety alongside it. It's crucial that we find a way to tap into a sense of mystery and awe if we're going to beat a path back toward each other—and feel a part of something larger and all-encompassing.

DIVINE PLAY

To create is to find a part of yourself that is often obscured in adulthood. By the time we've secured jobs and mortgages and built a family, creativity drops low on our list of priorities, becoming a luxury that we think we can no longer afford. But by engaging in creative acts, we are satisfying a fundamental human need to "enrich life," to use the words of Marshall Rosenberg.[11] And when we're not taking what is *alive* in us and making it manifest, this creates the true anxiety of the unexpressed creative impulse.

Creativity spans beyond painting or dancing to include any activity that elicits your own personal and unpredictable expression of self; it is an act that should be pursued for its own sake. It offers an essential sense of freedom, authenticity, and liveliness. Though we have increasingly deemphasized creativity and play in schools and in our culture—coinciding with the rising focus on testing and preparation for college—research has consistently shown that these pursuits are crucial to our well-being throughout life. For children, free play is necessary to finish the intricate wiring process of neural development;[12] in addition, researchers argue that children deprived of play are more likely to be anxious or depressed.[13,14] Throughout life, we need creativity and play to let our guard down, to give ourselves a sense of endless time and a wild sense of abandon.

Similarly, it's important to foster our connection to pleasure, whether that's through orgasm or opera or chocolate. Connecting to sexual pleasure, for example, is a way to tap into our life-giving energy. Physiologically speaking, sexual pleasure offers an antidote to anxiety. Orgasm prompts the body to release oxytocin,[15] our bonding hormone, as well as other feel-good hormones, such as dopamine.[16] These hormones have mood-boosting and antianxiety effects. And, perhaps most critically, those same hormones can promote feelings of connection.

NATURE

> Wildness is a necessity.
>
> —*John Muir*

We did not evolve in cubicles or factories, SUVs or subways, subterranean studios or gyms—we evolved in nature, with all its attendant sights, sounds, and smells. The need for natural surroundings is in our wiring, so being disconnected from nature makes us feel far from "home" and, in a profound way, not OK.

In Japan, there is a traditional practice called "forest bathing," thought to be an antidote to stress and various states of poor mental and physical health. Increasingly supported by the evidence, this practice of immersing oneself in nature is shown to reduce cortisol and improve mood and anxiety.[17,18] Neuroscientists have uncovered that walking through nature reduces rumination and can impact activity in the prefrontal cortex,[19] a part of the brain keenly involved in anxiety disorders.[20] There are even small studies that suggest that "earthing"—which includes such activities as walking barefoot on dirt or grass—may be healing, proposing that it allows a beneficial electron exchange to occur between your body and the earth.[21] You can choose to hike or climb or surf, or simply be still, finding your cathedral in the trees. These acts send a familiar signal to your brain that everything is all right—*you're home.*

As we discussed in chapter 5, spending time in nature is also the best way to reset your circadian rhythm. If your anxiety is causing nighttime disturbances (or insomnia is contributing to your anxiety), as with my patient Travis, getting back to our original contextual cues is one of the best ways to get sleep back on track.

Take the Plunge

I love to approximate "evolutionary conditions" as much as possible in modern life. I try to eat food that my body recognizes, avoid blue-spectrum light after sunset, and sleep in a cold room. I do, however, draw the line at one thing: hot showers. I'm not willing to sacrifice that oh-so-modern and unnatural luxury. And yet plunging into freezing cold water must have been a major part of our evolutionary conditions; research has shown that cold-water plunging decreases inflammation[22] and stimulates parasympathetic activity,[23] which has the potential to complete a stress cycle or establish a new, calmer baseline for the autonomic nervous system. I also find that, in the same way that psychedelics can offer seven years of therapy in one night, meditating in an ice bath can feel like three years of yoga in three minutes. That is, the body initially reacts to the cold water with tension, resistance, and stress; but if you can breathe, remain calm, and surrender in the face of the overwhelming sensations, in time, this develops muscle memory for maintaining equanimity when your body and mind are about to spiral toward panic.

So, if your progress has plateaued, you may want to revisit our evolutionary roots and investigate taking the plunge. Many people feel invigorated, relaxed, and less anxious once they get into the habit of daily cold showers or even—the horror!—ice baths. The Wim Hof Method[24] has formalized this practice, adding meditative breathing techniques to make the experience not just bearable but, for some, transformative.

"LIVE THE QUESTIONS"

For some people, the concept of source includes the idea of God. Perhaps at some point you had a deeply felt experience of the divine, and now this belief has become undeniable to you. Or you might feel so thoroughly convinced by scientific inquiry that the existence of a God is an impossible notion. Many of us live somewhere in between, struggling to feel certain in one direction or the other. "It's interesting to discover how many of [the great scientists I know] have deep religious faith despite the fact that it's been posited that somehow religion is contrary to science," President Barack Obama once mused in an interview, "but they don't see the contradiction."[25]

When I was growing up in the suburbs of New York City in the 1980s and '90s, the consensus around me regarding belief in God could be summed up as follows: believing in God was something you heard about on the country music station (before quickly turning the dial), and it was tacitly looked down upon. The religion around me was essentially scientism—that is, what we believed in and worshipped was scientific inquiry. Skepticism was a virtue.

I am all for critical thinking, but at the same time, I wonder if the intellectual elitism of science overlooked the true benefits of belief. Spiritual practice and congregating in the name of worship gives people community and a way of making sense of existence. Religion provides many people with a structure for regular spiritual inquiry. In an effort to buck the constraints and sometimes misguided beliefs of organized religion, many of us also lost our weekly ritual of communal seeking and asking the bigger questions of life. Spirituality is of course not the only way to add community, compassion, and purpose to our lives, but it's a big player in the landscape of how humans find connection and meaning.

To this day, I continue to love science. At the heart of science is the pursuit of truth and understanding how things work. However, some of the virtues of scientism might be making us anxious. In this worldview, everything we hold dear feels as though it's dangling at the mercy of randomness and chance. And scientism posits that we should meet everything with skepticism. Should we find ourselves enjoying a moment of awe and wonder at the mysteries of the universe, we must immediately sober up and point to the scientific explanation.

Here's the way to keep our pursuit of truth and science strong while also mitigating the background anxiety that comes with it: it's a both/and proposition. We are allowed to have one foot on the ground, rooted in science, and one foot floating in mystery. Perhaps the universe is both scientifically explainable *and* slightly magical. Maybe, in a way that's beyond our comprehension, both understandings of the world are simultaneously true.

Ultimately, we won't know who is right and who is wrong until we die and go to the room where they give us the answers. Or maybe there is no "room," and there's your answer. Either way, until one or the other happens, we have no way of truly knowing. So, in the meantime, like Obama's friends, despite my firm belief in the rigors of science, I've also decided to accept the idea of divine mystery.

If you're feeling isolated, alone, lost, and ungrounded, check in with yourself: Is there anywhere that you do feel a connection to something vast or far beyond the world that meets the eye? It could be when you're stargazing or deep in prayer, singing in a choir, or sitting in physics class. Whatever lights that spark for you, do that. Bring humility to these moments, knowing that there are benefits to exploring these mysteries with ourselves and in the company of others. We can "live the questions"—as Austrian poet Rainer Maria Rilke (1875–1926) puts it—as a powerful antidote to anxiety. If you threw out religion decades ago in a defiant cleanse of your conditioning, but you're feeling a hole where faith used to be, consider reclaiming a connection to spirituality and communal seeking in a way that feels good and true for you.

Chapter 15

Holding On, Letting Go

Trust in Allah but tie your camel.

—Ancient Arab adage, as interpreted by the scholar
Al-Tirmidhi, describing advice the prophet Muhammad
gave to a faithful Bedouin whom he saw leaving his
camel untethered

Your true anxiety is your ally, your North Star reliably guiding you. But once you've put its message into action, you can't just put your feet up and let the cosmos lead the way. On the contrary, even when you're following the directions of your deepest instincts, you'll still need to listen carefully for the next cue. Think of true anxiety as a kind of invisible electric fence, zapping you back on course every time you wander from your intended path. And when you're back on your path, that anxiety will transmute into a feeling of purpose.

All we ever really need to hope for, in fact, is to be on our path. It means we get to be ourselves, do what lights us up, and make a contribution that we are perfectly suited to make. When we befriend our true anxiety and let it guide us, it can serve us and the world. For some people this is grand, for others it happens on a less imposing scale, but it's never small—as long as we are being our authentic selves and making our unique offering, it's infinitely impactful. We operate with clarity and lift those around us. We feel guided, full of purpose, awake, and fulfilled.

LETTING GO

Recently, a patient of mine—a forty-two-year-old man I'll call Vincent—shared with me how letting go of control helped him overcome his anxiety. "All of this was really a quest for safety out of a fear of abandonment," he said, referring to his high-powered job in law as well as his many acquisitions, large and small, from real estate to fashionable clothes, "but the real answer is that there is no safety with a lowercase *s*—that safety is an illusion. And the real safety, the big *S* Safety, is *everywhere*. And it is *not* fragile. But transitioning from safety to Safety is very hard." Vincent had invested so much energy in securing his safety—the illusory sense of protection—that now it felt quite disorienting to not have to work so hard anymore, to simply rest in his trust of the larger Safety.

Letting go of the notion of control or surrendering to the larger forces at play—whatever your sense of what those forces may be—isn't easy. It is counterintuitive, and it can feel, at first, as if you're in free fall. As Sarah Wilson paraphrases American Tibetan Buddhist, ordained nun, and author Pema Chödrön in *First, We Make the Beast Beautiful*, anxiety is "resisting the unknown."[1]

We tend to white-knuckle our way through life, blaming ourselves and others when we don't get what we want. We're anxious and exhausted because we are fighting with reality, believing things are supposed to go a certain way. Instead of showing us where we need more control, anxiety actually alerts us to when we need to let go; when we need to take a breath and patiently, courageously see where our particular path will take us.

There's an old Taoist parable that illustrates how we can be too quick to judge what—of everything that happens to us in this life—is good or bad. As the story goes, a rice farmer has an old, sick horse. Bad, right? But wait, the farmer decides to let the horse go to spend its last days free in the mountain pastures. While the farmer's neighbors all agree this is clearly terrible news—the family has lost their horse—the farmer is not so certain. He says, "We'll see." Some weeks later, the horse returns much rejuvenated, along with a wild horse that has followed the old horse home. The neighbors want to congratulate the farmer, who suddenly went from no horses to two, for his good fortune, but the farmer pauses. "We'll see," he says again. When the farmer's only son tries to ride the new horse, he is thrown, and his leg is broken. *Now* would the farmer agree on his bad fortune? He resets his son's leg, and while it's still healing, the emperor declares a war. All able-bodied men in the village are drafted, and eventually all die in this war. Only the farmer's young son was spared, because he had a broken leg.

The lesson is not as all-encompassing or trite as "everything happens for a reason." Terrible things happen. Senseless violence, tragic accidents, abhorrent injustices, wildfires, and pandemics happen. And most of the world's tragedies disproportionately impact already vulnerable and marginalized populations. But while remaining resolute in fighting to make things right, it also behooves us to stay open to the unexpected twists and turns in the road ahead, finding whatever seed of peace we can in the midst of it all. Even if sometimes this means courageously sitting in the pain, patiently waiting to see what that seed becomes. We don't ever really know the ultimate significance of what is happening, so the more we can get comfortable with—or at least curious about—surrendering and trusting in something larger unfurling, the less anxious we'll be. When I'm tempted to resist reality, I remind myself to surrender to that which I can't control. Which, admittedly, is considerably more difficult when the loss is profound.

ON GRIEF

In July 2015, when I was six months pregnant with my daughter, my mother died suddenly. After all those years waiting patiently for grandkids, I wanted her to be able to enjoy this experience. I wanted her to know my daughter. I wanted my mom's gentle, grounding presence in my life as a mother. I *still* want this.

The loss of my mother steeply challenged my spiritual worldview. Just when I needed to believe in *something, anything,* I experienced crushing doubt. I wanted so desperately to know that my mom continued to exist in spirit form, that she would somehow have a way of knowing my daughter. I went looking for a rabbi, a priest, a shaman—honestly any grown-up would do—who could assure me that there was some greater order to the universe and that my mom's death was not a cruel act of chance.

My mother died too young, leaving me lonely, sad, longing for her to help me know how to mother my own child, looking on at my bereft father, unsure how to console him. At a certain point in my grieving, I realized that I had two options: One was to believe that death is senseless. The other was to believe that maybe there is more to it, and maybe I get to decide what it means for me. In the face of doubt, I chose to see my loss as part of an interconnected web of everything; this belief became woven into my understanding of some divine order. I still choose to see such events in this way. I'm aware that I could be deluded, but this view allows me to find meaning and comfort in challenging moments, rather than be left with bottomless anxiety and despair. It offers me a way of feeling just as close to my mother today as I did when she was alive. And, ultimately, it was only in surrendering to the fact of what had happened—radically accepting the death of my mother—that I was able to find, and keep finding, the calm at the center of it, the love and strength and calm in *me*.

So, if and when you get that phone call and your life is abruptly dropped into a worst-case-scenario moment, while you're swimming through the waves of surreality, anxiety, and pain, try to create a through line. If it feels available to you—even if it is just dimly beckoning to you—explore the possibility of order beyond your comprehension. The vicissitudes of our lives may indeed be senseless, but if you can find meaning in the unfolding events of your life, particularly the hard ones, this may help you move through them with more resilience and peace. If you can stay open to the possibility that we continue to exist in ways that are impossible to understand, that the mystery doesn't have to be an empty dream, then separation feels less permanent; loss feels less absolute.

Most of all, in moments of grief or challenge, be present. Stay wide awake. If it hurts, let it hurt. Let it hurt as much as it does, which can be so, so much. "When Grief comes to visit me, it's like being visited by a tsunami," the author Elizabeth Gilbert once remarked. "I am given just enough warning to say, 'Oh my god, this is happening RIGHT NOW,' and then I drop to the floor on my knees and let it rock me."[2] But she also later told Oprah, commenting on the agony of losing her partner to cancer: "I'll take it, I'll take all of it, I'll take the whole thing because I don't want to miss it. I don't want to have come all this distance to live a human life and miss the experience, so I just want to show up for the whole ride, whatever it might be."[3] Pushing away emotional pain does not erase it, but merely buries it in our bodies, where it can transmute into physical pain, illness, numbness, and rage—expressions of an incomplete *grief* cycle. Better to feel the full force of your emotions as they arise. You can handle it. You *want* to feel it. It's a way of honoring what you've lost.

My daughter is now six, and my mother is still, in a sense, with us. I believe she talks to me through particular songs, she visits me in dreams, and there are certain moments where I'm overcome by a tingling sensation that I equate with her presence. I can hear her and feel her—with what the Buddhist monk Sayadaw U. Pandita once called "a heart that is ready for anything"[4]—because I've chosen to listen.

YOUR TRUE PATH

If your soul is on fire to serve in a way that
scares you to death. And you don't feel worthy,
ready, able. And the burning will not leave you alone.
Congratulations. You have your calling.

—*Jaiya John*, Freedom: Medicine Words for
Your Brave Revolution

I've worked with my patient Valentina for years. At the age of forty-two, she's been through a great deal. Her father died when she was a toddler, and her childhood was anything but idyllic. Her mother, who didn't have the emotional wherewithal to process her grief or support Valentina through hers, developed a drinking problem while working two jobs and taking care of three kids alone. Valentina herself has had problems developing healthy romantic relationships, and she struggles with financial issues. One day, as we sat across from each other in my office, she talked about her Catholic faith and said she couldn't sign on for the idea that her father was sitting in Heaven looking down on her. "How could he be?" she asked. "I was two when he died. He doesn't even *know* me." I felt particularly struck by this because *my* daughter was two years old at the time, and I realized Valentina had it turned around. She didn't get to know her father (or at least, she couldn't consciously access memories of him because she was too young when he died)—a sad truth—but *he* almost certainly knew *her*. My daughter is everything to me and my husband, and even when she was barely a toddler, we were keenly aware of her character, personality, and spirit. I shared this with Valentina, and with that shift in perspective, she was able to imagine that her dad was actually out there somewhere, rooting for her, having known her before she even really knew herself. This was Valentina finding her path—the way she could imagine herself in connection with the universe—guided by love from some place beyond.

Once we develop the ability to listen to our body's inner whisper, we have the benefit of an internal compass telling us when we're moving in the right direction and when we've strayed. And that is all we ever need to know. We can never know what the future holds, and there is wisdom in being unattached to knowing. But as long as we have the conviction that we're where we need to be, we can loosen our grip. And if we know that we can trust ourselves, do we really still feel the need to control the outcome? Knowing we're on our path, headed in the right direction, is enough.

It takes conscious, daily awareness and effort to make sure our habits, work, and interactions with others align with our highest self. To me, the word *path* conveys this sense of both meaning and journey, direction and ongoingness that add up to a life that is more than the sum of its parts. *Path* also accentuates the notion that you don't have to have it all figured out; there is no pressure to arrive somewhere; it's simply the *direction* that needs to feel right.

If you are one of the sensitive ones—one of the artists, feelers, overthinkers, or intuitives of our human tribe—there will be some anxiety along the way, because as long as the world is imperfect, some truths are going to hurt, and you may feel these in a more visceral way than others do. Yours is a harder path, yet it's also a high calling. But honesty, and the courage to face each of our unique challenges, *can* shed light along the way. The work begins with stripping away what's blocking you from living in a state of physical balance, and it continues as you come to have faith in your true anxiety to keep you pointed in the right direction. When you fall out of alignment, your true anxiety will be there to nudge you back on course. Your path may not feel consequential, but it is. It is your highest expression; it is what your deepest anxiety has been insisting you follow all along.

Acknowledgments

Thank you, Vimal, for your patience, support, and immeasurable help, but most of all, thank you for your warmth, your heart, your humor, and your love. You are my infinity.

Thank you, J, for being my reason to work quickly, and for being a daily miracle. In the words of Brené Brown, "I will not teach or love or show you anything perfectly, but I will let you see me, and I will always hold sacred the gift of seeing you. Truly, deeply, seeing you."

Thank you, Mom. For everything. I miss you. I hope this book makes you proud.

Thank you to my patients. You are my greatest teachers, and it is a privilege to walk this path with you. And thank you for allowing me to share your stories with readers so they can find their path and feel less alone in the process.

Thank you, Nell Casey, for the immense care and skill you brought to every page of this work. To be two Nells, the daughters of two Janes, felt like just the right amount of synchronicity to know we were both exactly where we needed to be.

Thank you, Julie Will, for seeing the potential of this book, for your insights and savvy, and for bringing so much care to this project through a hectic and heavy time of life.

Thank you, John Maas and Celeste Fine. I felt that I was in

expert and caring hands throughout this process. And you both make it fun.

Thank you, Emma Kupor, for being an enormous help at every step of this process. I'm very grateful for all that you do.

Thank you, Mia Vitale, Sarah Passick, and the whole team at Park & Fine Literary, for taking such good care of me through this process.

Thank you, Anne Gerson. When the world felt like an abyss, you stepped in and filled it with all the warmth, knowledge, and traditions of our mom. Thank goodness for you, my brilliant sister.

Thank you, Dad, for making me feel so parented, even when we were all reeling. Thank you for your support and for being a source of stability and a model of steady, present parenting. Thank you so, so much. I love you.

Thank you, Nayana Vora, for stepping in and helping me feel mothered, and for showering J in the sweetest nectar of your embrace. You are pure love.

Thank you, Ashok Vora, Manish Vora, Ankur Vora, and Nisha Vora, for surrounding me with laughter, joy, card games, Ping-Pong, and love. I am so grateful to have you as my family.

Thank you, Adam Gerson, for a critical late-night editing session, for being an amazing partner to my sister, for being a parenting role model, and for being a true brother.

Thank you, Omri Navot, for creative brainstorming, for life shamanism, and for twenty-four years of deep friendship.

Thank you, Spencer Mash and Melissa Shin Mash, for being our chosen family and our village.

Thank you, Chris Moreno and Karen Appelquist. Your friendship fills life with laughter and joy, and "V Work" has kept me on track (or maybe it actually distracted me, but it was well worth it). Here's to many future adventures.

Thank you, Jennifer Drapkin, for your friendship, humor, brilliance, and magic.

Thank you, Cat Loerke and Carl Erik Fisher. You are both brilliant and you have your fingers on the pulse. I couldn't ask for a better brain trust. I learn so much from both of you, and I take so much comfort in our friendship.

Thank you, Sarah Messmore, for your quietly brilliant insights and for always helping me have faith in myself when things feel heavy. Your students are so lucky to have you in their lives.

Thank you, Jeremy Ortman, Stephen Sosnowski, and Tamar Steinberger, for a very enjoyable sun-soaked brainstorming session. J'accuse!

Thank you, Teenisha Toussaint, for sharing your gifts with our children and for really, truly seeing J. Our year with you shifted our entire trajectory.

Thank you, Jenny and Ken Young, for creating the most incredible pod, which allowed J to thrive and gave me space to write during an immensely challenging year. I hope to always be on the same spaceship as the two of you.

Thank you, Paul Kuhn and Erica Matluck, for your friendship, wisdom, music, and endless inspiration. You are both shimmering souls illuminating everything you touch. Let's live this next phase together.

Thank you, Bing Cheah and Navlyn Wang, for being our kindred spirit seekers. You teach us about joy, self-reflection, and divine play every time we connect.

Thank you, Seanna Sifflet and Brady Ovson, for a decade of the warmest friendship and the most irresponsible but worthwhile two a.m. hangouts. Love you both so, so much.

Thank you, Maryellis Bunn, for being the most incredible auntie and for being an inspiration and a beacon of growth.

Thank you, Robin Marie Younkin. Your contributions to this book were indispensable, and you make my entire professional life possible with your calm, steady presence.

Thank you, Stephanie Higgs. You helped me understand the essential structure to tell this story, and your beautiful turns of phrase live on throughout this book.

Thank you, Melissa Urban. You are a light, an inspiration, a gift to the world, and a good friend.

Thank you, Holly Whitaker, for friendship, inspiration, and permission to trust this unfolding.

Thank you, Will Cole, DC, for being the most grounded heart of gold in the wellness world.

Thank you, Will DeRooy, for your meticulous and efficient work. I'm grateful to have been in such good hands.

Thank you, Jeremy Fisher, for helping a very important section come to life.

Thank you, Priya Ahuja, for generously sharing your deep knowledge of Chinese medicine.

Thank you, Chris Kresser, for being my first mentor in the functional medicine space.

Thank you, Ron Rieder, MD, for giving me a long enough leash to pursue passions beyond the conventional.

Thank you, Frank Lipman, MD, for encouraging me to spread my ideas beyond my practice.

Thank you, Tom Lee, MD, for giving me so many venues to find my voice and pioneer new ideas for health and healing.

Thank you, Jason Wachob and Colleen Wachob, for your kindness and generosity, and for being my first gateway to spreading my ideas on a larger stage.

Thank you, Andrew Chomer and Roya Darling, for the gift of aligned friendship. This is only the beginning. Let's set an intention and manifest it together.

Thank you, Barbara LaPine. The way you taught the subject of biology—with clarity, nuance, and passion—is the reason I pursued medicine.

Thank you, momtribe, for being the village that makes parenting possible.

Thank you, 305 crew, for endless laughter, and for keeping me honest and on my toes.

Appendix:
Herbs and Supplements
for Anxiety

As you have probably gathered by now, supplements are not the focus of my approach to anxiety. I believe the backbone of addressing anxiety comes down to diet and lifestyle modifications paired with psychospiritual healing. And I think the pill-for-an-ill idea of conventional medicine misses the point; just as I don't think anxiety is a Lexapro-deficiency disorder, it's not an L-theanine-deficiency disorder either. I will concede, however, that there are some vitamins, minerals, and herbs worth considering. This is not an exhaustive list, but these are the supplements I recommend from time to time in my practice.

- **Magnesium glycinate:** As discussed, it's hard to get this in sufficient quantities from food, so I think most of us should be supplementing with 100–800 mg of magnesium glycinate at bedtime to round out our nutritional needs. This can be helpful for anxiety, insomnia, menstrual cramps, and headaches. If you develop loose stool, take less.
- **Methylated B vitamins:**[1] Many people are depleted in the B vitamins, whether that's from inadequate nutrition, chronic

stress, or a side effect of the birth control pill. Many of my patients also have a common genetic variation called the MTHFR mutation, which can make it difficult for their bodies to methylate folate—essentially a compromised ability to convert some of the B vitamins into their active form. For this reason, I find it generally useful to supplement with premethylated B vitamins, such as L-methylfolate and methylcobalamin. Methylcobalamin (i.e., vitamin B_{12}) is particularly important for anyone following a vegetarian or vegan diet.

- **Vitamin D:** As discussed, I prefer that people get their vitamin D from the sun, but when that's not possible or safe, a little supervised supplementation can be a helpful support. Taking vitamin D_3 alongside magnesium and vitamins A, E, and K_2 is a particularly beneficial combination.

- **Curcumin/turmeric:** When inflammation is playing a role in anxiety, this is a helpful support to get a dysregulated immune system back into balance. Curcumin becomes more bioavailable when combined with black pepper and ghee.

- **Cod liver oil:** This is not an anxiety supplement per se, but as a source of omega-3 fatty acids, I find it to be part of an overall strategy for ensuring healthy neuronal cell membranes. I prefer cod liver oil to fish oil because cod liver oil also provides the fat-soluble vitamins D, A, E, and K.

- **CBD oil/hemp oil:** Many of my patients have dabbled in hemp oil for a few days only to declare it ineffective. Please note that it's often necessary to use hemp oil consistently for several weeks to experience the full calming effect.

- **L-theanine:** A constituent of green tea, some of my patients find this to be a helpful support for their anxiety, and there is research to back this up.[2]

- **Bach Rescue Remedy:** This is a flower essence that some of my patients like to take "as needed" for anxiety.
- **NAC (N-acetyl-cysteine):** This is a precursor to glutathione, the most important antioxidant in the body. I find NAC particularly helpful for patients going through medication discontinuation. I also find it helpful for bipolar support.
- **Ashwagandha:** An adaptogenic herb from the Ayurvedic tradition, ashwagandha has been helpful for some of my patients, especially women, to take right before bed. Ashwagandha, however, is contraindicated during pregnancy and lactation.[3]
- **Inositol:** When inositol is indicated, I generally recommend mixing powdered myoinositol (the most bioavailable form of inositol) with water in a large water bottle, carrying it around with you, and sipping it throughout the day. You will want to start with a small amount and increase gradually to 18 grams of myoinositol daily, as it can cause GI distress when started too quickly. Do a trial for four to six weeks and then pause. I find this particularly helpful for OCD. It can also be effective for depression and panic disorder.[4,5]
- **Passionflower:** Some of my patients have found this especially helpful for cognitive anxiety (when they are, for example, stuck in a ruminative spiral). Note, however, that passionflower is contraindicated with SSRIs.
- **Chamomile and tulsi:** I often recommend a chamomile or tulsi tea ritual in the evening to help patients relax before bedtime.
- **Probiotics:** I generally recommend supporting gut health with fermented foods rather than probiotics, though there are times—such as when repleting the gut flora after a

course of antibiotics or after a small intestinal bacterial overgrowth (SIBO) treatment under the supervision of a naturopath or functional medicine practitioner—when probiotics are a necessary addition.

- **Phosphatidylserine:** Some find this to be a helpful salve to treat the effects of chronic stress.
- **Skullcap, milky oats, and lemon balm:** Some of my patients find these to be especially helpful for treating their body-based anxiety (e.g., a physiological stress response).
- **Cannabis:** I've had patients for whom cannabis was a good and safe medicine—helping them with anxiety, insomnia, menstrual cramps, and benzodiazepine dependence.[6] And I've had patients whose cannabis dependence held them back in their lives. The takeaway: it's worth a discussion with a knowledgeable provider to understand the balance of risks and benefits for you.

Notes

Chapter 1: The Age of Anxiety

1. Ruscio, A. M., Hallion, L. S., Lim, C., Aguilar-Gaxiola, S., Al-Hamzawi, A., Alonso, J., Andrade, L. H., Borges, G., Bromet, E. J., Bunting, B., Caldas de Almeida, J. M., Demyttenaere, K., Florescu, S., de Girolamo, G., Gureje, O., Haro, J. M., He, Y., Hinkov, H., Hu, C., de Jonge, P., Scott, K. M., et al. (2017). "Cross-Sectional Comparison of the Epidemiology of DSM-5 Generalized Anxiety Disorder across the Globe." *JAMA Psychiatry* 74 (5): 465–475. https://doi.org/10.1001/jamapsychiatry.2017.0056.

2. Bandelow, B., & Michaelis, S. (2015). "Epidemiology of Anxiety Disorders in the 21st Century." *Dialogues in Clinical Neuroscience* 17 (3): 327–335. https://doi.org/10.31887/dcns.2015.17.3/bbandelow.

3. Goodwin, R. D., Weinberger, A. H., Kim, J. H., Wu, M., & Galea, S. (2020). "Trends in Anxiety among Adults in the United States, 2008–2018: Rapid Increases among Young Adults." *Journal of Psychiatric Research* 130: 441–446. https://doi.org/10.1016/j.jpsychires.2020.08.014.

4. Pancha, N., Kamal, R., Cox, C., & Garfield, R. (2021). "The Implications of COVID-19 for Mental Health and Substance Use." KFF, February 10. www.kff.org/coronavirus-covid-19/issue-brief/the-implications-of-covid-19-for-mental-health-and-substance-use/.

5. Crocq, M.-A. (2015). "A History of Anxiety: From Hippocrates to DSM." *Dialogues in Clinical Neuroscience* 17 (3): 319–325. https://doi.org/10.31887/DCNS.2015.17.3/macrocq.

6. Crocq, "A History of Anxiety."

7. Crocq, "A History of Anxiety."

8. Ross, J. (2002). *The Mood Cure: The 4-Step Program to Rebalance Your Emotional Chemistry and Rediscover Your Natural Sense of Well-Being* (New York: Viking), 4.

Chapter 2: Avoidable Anxiety

1. Jacka, F. N., O'Neil, A., Opie, R., Itsiopoulos, C., Cotton, S., Mohebbi, M., Castle, D., Dash, S., Mihalopoulos, C., Chatterton, M. L., Brazionis, L., Dean, O. M., Hodge, A. M., & Berk, M. (2017). "A Randomised Controlled Trial of Dietary Improvement for Adults with Major Depression (the 'SMILES' Trial)." *BMC Medicine* 15 (1): 23. https://doi.org/10.1186/s12916-017-0791-y.

2. Ramaholimihaso, T., Bouazzaoui, F., & Kaladjian, A. (2020). "Curcumin in Depression: Potential Mechanisms of Action and Current Evidence—A Narrative Review." *Frontiers in Psychiatry* 11. https://doi.org/10.3389/fpsyt.2020.572533.

3. Nollet, M., Wisden, W., & Franks, N. (2020). "Sleep Deprivation and Stress: A Reciprocal Relationship." *Interface Focus* 10 (3). https://doi.org/10.1098/rsfs.2019.0092. .

4. Lovallo, W., Whitsett, T., al'Absi, M., Sung, B., Vincent, A., & Wilson, M. (2005). "Caffeine Stimulation of Cortisol Secretion across the Waking Hours in Relation to Caffeine Intake Levels." *Psychosomatic Medicine* 67 (5): 734–739. https://doi.org/10.1097/01.psy.0000181270.20036.06.

5. Nagoski, E., & Nagoski, A. (2019). *Burnout: The Secret to Unlocking the Stress Cycle* (New York: Ballantine), 15.

6. Vighi, G., Marcucci, F., Sensi, L., Di Cara, G., & Frati, F. (2008). "Allergy and the Gastrointestinal System." Supplement, *Clinical and Experimental Immunology* 153 (S1): 3–6. https://doi.org/10.1111/j.1365-2249.2008.03713.x.

7. Hadhazy, A. (2010). "Think Twice: How the Gut's 'Second Brain' Influences Mood and Well-Being." *Scientific American*, February 12. www.scientificamerican.com/article/gut-second-brain.

8. Breit, Sigrid, et al. (2018). "Vagus Nerve as Modulator of the Brain–Gut Axis in Psychiatric and Inflammatory Disorders." *Frontiers in Psychiatry* 9: 44. https://dx.doi.org/10.3389%2Ffpsyt.2018.00044.

9. Pokusaeva, K., Johnson, C., Luk, B., Uribe, G., Fu, Y., Oezguen, N., Matsunami, R. K., et al. (2016). "GABA-Producing *Bifidobacterium dentium* Modulates Visceral Sensitivity in the Intestine." *Neurogastroenterology & Motility* 29 (1). https://doi.org /10.1111/nmo.12904.

10. Strandwitz, P., Kim, K. H., Terekhova, D., Liu, J. K., Sharma, A., Levering, J., McDonald, D., et al. (2018). "GABA-Modulating Bacteria of the Human Gut Microbiota." *Nature Microbiology* 4: 396–403. https://doi.org/10.1038/s41564-018 -0307-3.

11. Clapp, M., Aurora, N., Herrera, L., Bhatia, M., Wilen, E., & Wakefield, S. (2017). "Gut Microbiota's Effect on Mental Health: The Gut-Brain Axis." *Clinics and Practice* 7 (4): 987. https://doi.org/10.4081/cp.2017.987.

12. Cooper, P. J. (2009). "Interactions between Helminth Parasites and Allergy." *Current Opinion in Allergy and Clinical Immunology* 9 (1): 29–37. https://doi.org/10.1097/ACI.0b013e 32831f44a6.

Chapter 3: Purposeful Anxiety

1. Moody, L., in conversation with Glennon Doyle. (2020). "Glennon Doyle on Overcoming Lyme Disease, Hope During Hard Times, and the Best Relationship Advice." *Healthier Together* (podcast), https://www.lizmoody.com/healthiertogether podcast-glennon-doyle.

2. Wilson, Sarah. (2018). *First, We Make the Beast Beautiful: A New Journey through Anxiety* (New York: Dey Street), 164.

3. Fitzgerald, F. Scott. (1936). "The Crack-Up." *Esquire*, February.

Chapter 5: Tired and Wired

1. Anxiety and Depression Association of America. (2021). "Sleep Disorders." https://adaa.org/understanding-anxiety/related -illnesses/sleep-disorders.

2. Rasch, B., & Born, J. (2013). "About Sleep's Role in Memory." *Physiological Reviews* 93 (2): 681–766. https://doi.org/10.1152 /physrev.00032.2012.

3. Eugene, A. R., & Masiak, J. (2015). "The Neuroprotective Aspects of Sleep." *MEDtube Science* 3 (1): 35–40. https:// pubmed.ncbi.nlm.nih.gov/26594659.

4. Dimitrov, S., Lange, T., Gouttefangeas, C., Jensen, A., Szczepanski, M., Lehnnolz, J., Soekadar, S., et al. (2019). "Gαs-Coupled Receptor Signaling and Sleep Regulate Integrin Activation of Human Antigen-Specific T Cells." *Journal of Experimental Medicine* 216 (3): 517–526. https://doi.org/10.1084/jem.20181169.

5. Nunez, K., & Lamoreux, K. (2020). "What Is the Purpose of Sleep?" Healthline, July 20. www.healthline.com/health/why-do-we-sleep.

6. Scharf, M. T., Naidoo, N., Zimmerman, J. E., & Pack, A. I. (2008). "The Energy Hypothesis of Sleep Revisited." *Progress in Neurobiology* 86 (3): 264–280. https://doi.org/10.1016/j.pneurobio.2008.08.003.

7. Jessen, N. A., Munk, A. S., Lundgaard, I., & Nedergaard, M. (2015). "The Glymphatic System: A Beginner's Guide." *Neurochemical Research* 40 (12): 2583–2599. https://doi.org/10.1007/s11064-015-1581-6.

8. Xie, L., Kang, H., Xu, Q., Chen, M. J., Liao, Y., Thiyagarajan, M., O'Donnell, J., Christensen, D. J., Nicholson, C., Iliff, J. J., Takano, T., Deane, R., & Nedergaard, M. (2013). "Sleep Drives Metabolite Clearance from the Adult Brain." *Science* 342 (6156): 373–377. https://doi.org/10.1126/science.1241224.

9. Benveniste, H., Liu, X., Koundal, S., Sanggaard, S., Lee, H., & Wardlaw, J. (2019). "The Glymphatic System and Waste Clearance with Brain Aging: A Review." *Gerontology* 65 (2): 106–119. https://doi.org/10.1159/000490349.

10. Xie et al., "Sleep Drives Metabolite Clearance from the Adult Brain."

11. Reddy, O. C., & van der Werf, Y. D. (2020). "The Sleeping Brain: Harnessing the Power of the Glymphatic System through Lifestyle Choices." *Brain Sciences* 10 (11): 868. https://doi.org/10.3390/brainsci10110868.

12. Tähkämö, L., Partonen, T., & Pesonen, A.-K. (2019). "Systematic Review of Light Exposure Impact on Human Circadian Rhythm." *Chronobiology International* 36 (2): 151–170. https://doi.org/10.1080/07420528.2018.1527773.

13. Peplonska, B., Bukowska, A., & Sobala, W. (2015). "Association of Rotating Night Shift Work with BMI and Abdominal

Obesity among Nurses and Midwives." *PLoS ONE* 10 (7). https://doi.org/10.1371/journal.pone.0133761.

14. Vetter, C., Devore, E. E., Wegrzyn, L. R., Massa, J., Speizer, F. E., Kawachi, I., Rosner, B., Stampfer, M. J., & Schernhammer, E. S. (2016). "Association between Rotating Night Shift Work and Risk of Coronary Heart Disease among Women." *JAMA* 315 (16): 1726–1734. https://doi.org/10.1001/jama.2016.4454.

15. Wegrzyn, L. R., Tamimi, R. M., Rosner, B. A., Brown, S. B., Stevens, R. G., Eliassen, A. H., Laden, F., Willett, W. C., Hankinson, S. E., & Schernhammer, E. S. (2017). "Rotating Night-Shift Work and the Risk of Breast Cancer in the Nurses' Health Studies." *American Journal of Epidemiology* 186 (5): 532–540. https://doi.org/10.1093/aje/kwx140.

16. Szkiela, M., Kusideł, E., Makowiec-Dąbrowska, T., & Kaleta, D. (2020). "Night Shift Work: A Risk Factor for Breast Cancer." *International Journal of Environmental Research and Public Health* 17 (2): 659. https://doi.org/10.3390/ijerph17020659.

17. Taheri, S., Lin, L., Austin, D., Young, T., & Mignot, E. (2004). "Short Sleep Duration Is Associated with Reduced Leptin, Elevated Ghrelin, and Increased Body Mass Index." *PLoS Medicine* 1 (3). https://doi.org/10.1371/journal.pmed.0010062.

18. Yetish, G., Kaplan, H., Gurven, M., Wood, B., Pontzer, H., Manger, P. R., Wilson, C., McGregor, R., & Siegel, J. M. (2015). "Natural Sleep and Its Seasonal Variations in Three Pre-industrial Societies." *Current Biology* 25 (21): 2862–2868. https://doi.org/10.1016/j.cub.2015.09.046.

19. Whitwell, T. (2020). "52 Things I Learned in 2020." Medium, December 1. https://medium.com/fluxx-studio-notes/52-things-i-learned-in-2020-6a380692dbb8.

20. Institute of Medicine (US) Committee on Military Nutrition Research. (2001). "Pharmacology of Caffeine," in *Caffeine for the Sustainment of Mental Task Performance: Formulations for Military Operations* (Washington, DC: National Academies Press). www.ncbi.nlm.nih.gov/books/NBK223808/.

21. Roenneberg, Till. (2012). *Internal Time: Chronotypes, Social Jet Lag, and Why You're So Tired* (Cambridge, MA: Harvard University Press).

22. He, Y., Jones, C. R., Fujiki, N., Xu, Y., Guo, B., Holder Jr., J. L., Rossner, M. J., Nishino, S., & Fu, Y. H. (2009). "The

Transcriptional Repressor DEC2 Regulates Sleep Length in Mammals." *Science* 325 (5942): 866–870. https://doi.org/10.1126/science.1174443.

23. Chaput, J. P., Dutil, C., & Sampasa-Kanyinga, H. (2018). "Sleeping Hours: What Is the Ideal Number and How Does Age Impact This?" *Nature and Science of Sleep* 10: 421–430. https://doi.org/10.2147/NSS.S163071.

24. Shi, G., Xing, L., Wu, D., Bhattacharyya, B. J., Jones, C. R., McMahon, T., Chong, S. Y. C., et al. (2019). "A Rare Mutation of β_1-Adrenergic Receptor Affects Sleep/Wake Behaviors." *Neuron* 103 (6): 1044–1055. https://doi.org/10.1016/j.neuron.2019.07.026.

25. Watson, N. F., Badr, M. S., Belenky, G., Bliwise, D. L., Buxton, O. M., Buysse, D., Dinges, D. F., Gangwisch, J., Grandner, M. A., Kushida, C., Malhotra, R. K., Martin, J. L., Patel, S. R., Quan, S. F., & Tasali, E. (2015). "Recommended Amount of Sleep for a Healthy Adult: A Joint Consensus Statement of the American Academy of Sleep Medicine and Sleep Research Society." *Sleep* 38 (6): 843–844. https://doi.org/10.5665/sleep.4716.

26. Scullin, M. K., Krueger, M. L., Ballard, H. K., Pruett, N., & Bliwise, D. L. (2018). "The Effects of Bedtime Writing on Difficulty Falling Asleep: A Polysomnographic Study Comparing To-Do Lists and Completed Activity Lists." *Journal of Experimental Psychology: General* 147 (1): 139–146. https://doi.org/10.1037/xge0000374.

27. Boyle, N. B., Lawton, C. L., & Dye, L. (2017). "The Effects of Magnesium Supplementation on Subjective Anxiety and Stress—a Systematic Review." *Nutrients* 9 (5): 429. https://doi.org/10.3390/nu9050429.

28. Serefko, A., Szopa, A., & Poleszak, E. (2016). "Magnesium and Depression." *Magnesium Research* 29 (3): 112–119. https://pubmed.ncbi.nlm.nih.gov/27910808.

29. Chiu, H. Y., Yeh, T.-H., Huang, Y.-C., & Chen, P.-Y. (2016). "Effects of Intravenous and Oral Magnesium on Reducing Migraine: A Meta-Analysis of Randomized Controlled Trials." *Pain Physician* 19 (1): E97–E112. https://pubmed.ncbi.nlm.nih.gov/26752497.

30. Parazzini, F., Di Martino, M., & Pellegrino, P. (2017). "Magnesium in the Gynecological Practice: A Literature Review."

Magnesium Research 30 (1): 1–7. https://doi.org/10.1684/mrh
.2017.0419.

31.   Eron, K., Kohnert, L., Watters, A., Logan, C., Weisner-Rose, M.,
& Mehler, P. S. (2020). "Weighted Blanket Use: A Systematic
Review." *AJOT: The American Journal of Occupational Therapy*
74 (2). https://ajot.aota.org/article.aspx?articleid=2763119.

32.   Onen, S. H., Onen, F., Bailly, D., & Parquet, P. (1994).
"Prévention et traitement des dyssomnies par une hygiène
du sommeil" [Prevention and Treatment of Sleep Disorders
through Regulation of Sleeping Habits]. *La Presse Médicale* 23
(10): 485–489. https://pubmed.ncbi.nlm.nih.gov/8022726.

33.   Ebrahim, I., Shapiro, C., Williams, A., & Fenwick, P. (2013).
"Alcohol and Sleep I: Effects on Normal Sleep." *Alcoholism:
Clinical and Experimental Research* 37 (4): 539–549. https://
doi.org/10.1111/acer.12006.

34.   Amaral, F. G., & Cipolla-Neto, J. (2018). "A Brief Review about
Melatonin, a Pineal Hormone." *Archives of Endocrinology and
Metabolism* 62 (4): 472–479. https://doi.org/10.20945/2359
-3997000000066.

35.   Cipolla-Neto, J., & Amaral, F. (2018). "Melatonin as a Hormone:
New Physiological and Clinical Insights." *Endocrine Reviews*
39 (6): 990–1028. https://doi.org/10.1210/er.2018-00084.

Chapter 6: Techxiety

1.   Haidt, J., & Twenge, J. (2019). "Social Media Use and
Mental Health: A Review." Unpublished manuscript, New
York University. https://docs.google.com/document/d/1w-
HOfseF2wF9YIpXwUUtP65-olnkPyWcgF5BiAtBEy0/edit#.

2.   Yuen, E. K., Koterba, E. A., Stasio, M., Patrick, R., Gangi, C.,
Ash, P., Barakat, K., Greene, V., Hamilton, W., & Mansour,
B. (2018). "The Effects of Facebook on Mood in Emerging
Adults." *Psychology of Popular Media Culture* 8 (3): 198–206.

3.   Shakya, H. B., & Christakis, N. A. (2017). "Association of
Facebook Use with Compromised Well-Being: A Longitudinal Study." *American Journal of Epidemiology* 185 (3): 203–211.
https://doi.org/10.1093/aje/kww189.

4.   Ducharme, J. (2021). "COVID-19 Is Making America's
Loneliness Epidemic Even Worse." *Time*, May 8. https://time
.com/5833681/loneliness-covid-19/.

5. Loades, M. E., Chatburn, E., Higson-Sweeney, N., Reynolds, S., Shafran, R., Brigden, A., Linney, C., McManus, M. N., Borwick, C., & Crawley, E. (2020). "Rapid Systematic Review: The Impact of Social Isolation and Loneliness on the Mental Health of Children and Adolescents in the Context of COVID-19." *Journal of the American Academy of Child and Adolescent Psychiatry* 59 (11): 1218–1239. https://doi.org/10.1016/j.jaac.2020.05.009.

6. Twenge, J. M., Cooper, A. B., Joiner, T. E., Duffy, M. E., & Binau, S. G. (2019). "Age, Period, and Cohort Trends in Mood Disorder Indicators and Suicide-Related Outcomes in a Nationally Representative Dataset, 2005–2017." *Journal of Abnormal Psychology* 128 (3): 185–199. https://doi.org/10.1037/abn0000410.

7. Twenge, J. M., Martin, G. N., & Spitzberg, B. H. (2019). "Trends in U.S. Adolescents' Media Use, 1976–2016: The Rise of Digital Media, the Decline of TV, and the (Near) Demise of Print." *Psychology of Popular Media Culture* 8 (4): 329–345. http://dx.doi.org/10.1037/ppm0000203.

8. Riehm, K. E., Feder, K. A., Tormohlen, K. N., Crum, R. M., Young, A. S., Green, K. M., Pacek, L. R., La Flair, L. N., & Mojtabai, R. (2019). "Associations between Time Spent Using Social Media and Internalizing and Externalizing Problems among US Youth." *JAMA Psychiatry* 76 (12): 1266–1273. https://doi.org/10.1001/jamapsychiatry.2019.2325.

9. Lukianoff, Greg, & Haidt, Jonathan. (2018). *The Coddling of the American Mind: How Good Intentions and Bad Ideas Are Setting Up a Generation for Failure* (New York: Penguin), 161.

10. Barthorpe, A., Winstone, L., Mars, B., & Moran, P. (2020). "Is Social Media Screen Time Really Associated with Poor Adolescent Mental Health? A Time Use Diary Study." *Journal of Affective Disorders* 274: 864–870. https://doi.org/10.1016/j.jad.2020.05.106.

11. Saeri, A. K, Cruwys, T., Barlow, F. K., Stronge, S., & Sibley, C. G. (2017). "Social Connectedness Improves Public Mental Health: Investigating Bidirectional Relationships in the New Zealand Attitudes and Values Survey." *Australian & New Zealand Journal of Psychiatry* 52 (4): 365–374. https://doi.org/10.1177/0004867417723990.

12. Saeri et al., "Social Connectedness."
13. Lieberman, Matthew D. (2013). *Social: Why Our Brains Are Wired to Connect* (Oxford: Oxford University Press), 9.
14. Wheeler, M. J., Dunstan, D. W., Smith, B., Smith, K. J., Scheer, A., Lewis, J., Naylor, L. H., Heinonen, I., Ellis, K. A., Cerin, E., Ainslie, P. N., & Green, D. J. (2019). "Morning Exercise Mitigates the Impact of Prolonged Sitting on Cerebral Blood Flow in Older Adults." *Journal of Applied Physiology* 126 (4): 1049–1055. https://doi.org/10.1152/japplphysiol.00001.2019.
15. Lee, Dave. (2017). "Facebook Founding President Sounds Alarm." BBC News, November 9. https://www.bbc.com/news/technology-41936791.
16. Börchers, Stina. (2021). "Your Brain on Instagram, TikTok, & Co.—The Neuroscience of Social Media." *Biologista* (blog), June 29. https://biologista.org/2020/06/29/your-brain-on-instagram-tiktok-co-the-neuroscience-of-social-media/.
17. Lee, "Facebook Founding President Sounds Alarm."
18. Tolle, Eckhart. (1999). *The Power of Now: A Guide to Spiritual Enlightenment* (Novato, CA: New World Library), 22.
19. Packnett Cunningham, Brittany N. (@MsPackyetti). "back . . . but barely!" Twitter, September 2, 2021, https://twitter.com/MsPackyetti/status/1433294762153496576.

Chapter 7: Food for Thought

1. Nestle, M. (1993). "Food Lobbies, the Food Pyramid, and U.S. Nutrition Policy." *International Journal of Health Services* 23 (3): 483–496. https://doi.org/10.2190/32f2-2pfb-meg7-8hpu.
2. Brown, Brené. (2019). "What Being Sober Has Meant to Me." *Brené Brown* (blog), May 31. https://brenebrown.com/blog/2019/05/31/what-being-sober-has-meant-to-me.
3. Fukudome, S., & Yoshikawa, M. (1992). "Opioid Peptides Derived from Wheat Gluten: Their Isolation and Characterization." *FEBS Letters* 296 (1): 107–111. https://doi.org/10.1016/0014-5793(92)80414-c.
4. Malav, T., Zhang, Y., Lopez-Toledano, M., Clarke, A., & Deth, R. (2016). "Differential Neurogenic Effects of Casein-Derived Opioid Peptides on Neuronal Stem Cells: Implications for Redox-Based Epigenetic Changes." *Journal of Nutritional Biochemistry* 37: 39–46. https://doi.org/10.1016/j.jnutbio.2015.10.012.

5. Ekren, Cansu. (2021). "Jameela Jamil Opens Up about Eating Disorder She Suffered from for Years." The Red Carpet, January 3. https://theredcarpet.net/jameela-jamil-opens-up-about-the-eating-disorder-she-experienced-for-years.

6. Blanco-Rojo, R., Sandoval-Insausti, H., López-Garcia, E., Graciani, A., Ordovás, J. M., Banegas, J. R., Rodríguez-Artalejo, F., & Guallar-Castillón, P. (2019). "Consumption of Ultra-Processed Foods and Mortality: A National Prospective Cohort in Spain." Mayo Clinic Proceedings 94 (11): 2178–2188. https://doi.org/10.1016/j.mayocp.2019.03.035.

7. Swaminathan, S., Dehghan, M., Raj, J. M., Thomas, T., Rangarajan, S., Jenkins, D., Mony, P., et al. (2021). "Associations of Cereal Grains Intake with Cardiovascular Disease and Mortality across 21 Countries in Prospective Urban and Rural Epidemiology Study: Prospective Cohort Study." BMJ 372: m4948. https://doi.org/10.1136/bmj.m4948.

8. Elizabeth, L., Machado, P., Zinöcker, M., Baker, P., & Lawrence, M. (2020). "Ultra-Processed Foods and Health Outcomes: A Narrative Review." Nutrients 12 (7): 1955. https://doi.org/10.3390/nu12071955.

9. O'Connor, A. (2016). "How the Sugar Industry Shifted Blame to Fat." New York Times, September 12. www.nytimes.com/2016/09/13/well/eat/how-the-sugar-industry-shifted-blame-to-fat.html.

10. Tesfaye, N., & Seaquist, E. R. (2010). "Neuroendocrine Responses to Hypoglycemia." Annals of the New York Academy of Sciences 1212 (1): 12–28. https://doi.org/10.1111/j.1749-6632.2010.05820.x.

11. Gonder-Frederick, L. A., Cox, D. J., Bobbitt, S. A., & Pennebaker, J. W. (1989). "Mood Changes Associated with Blood Glucose Fluctuations in Insulin-Dependent Diabetes Mellitus." Health Psychology 8 (1): 45–59. https://doi.org/10.1037//0278-6133.8.1.45.

12. Urban, M. (2019). "Taming Your Sugar Dragon, Part 1." Whole30 (blog), July 24. https://whole30.com/sugar-dragon-1. Revised per personal communication on March 30, 2021.

13. Alexander, Scott. (2015). "Things That Sometimes Work if You Have Anxiety." Slate Star Codex (blog), July 13. https://

slatestarcodex.com/2015/07/13/things-that-sometimes-work
-if-you-have-anxiety.

14. Ascherio, A., Zhang, S. M., Hernán, M. A., Kawachi, I.,
    Colditz, G. A., Speizer, F. E., & Willett, W. C. (2001). "Pro-
    spective Study of Caffeine Consumption and Risk of Parkin-
    son's Disease in Men and Women." *Annals of Neurology* 50 (1):
    56–63. https://doi.org/10.1002/ana.1052.

15. Moore, Charles. (2015). "Coffee Drinking Lowers Risk of
    Parkinson's, Type 2 Diabetes, Five Cancers, and More—Harvard
    Researchers." *Parkinson's News Today*, October 2. https://
    parkinsonsnewstoday.com/2015/10/02/coffee-drinking
    -lowers-risk-parkinsons-type-2-diabetes-five-cancers-harvard
    -researchers.

16. Lovallo, W. R., Whitsett, T. L., al'Absi, M., Sung, B. H., Vin-
    cent, A. S., & Wilson, M. F. (2005). "Caffeine Stimulation
    of Cortisol Secretion across the Waking Hours in Relation to
    Caffeine Intake Levels." *Psychosomatic Medicine* 67 (5): 734–
    739. https://doi.org/10.1097/01.psy.0000181270.20036.06.

17. Lane, J. D., & Williams Jr., R. B. (1987). "Cardiovascular
    Effects of Caffeine and Stress in Regular Coffee Drinkers."
    Psychophysiology 24 (2): 157–164. https://doi.org/10.1111/j.1469
    -8986.1987.tb00271.x.

18. Winston, A., Hardwick, E., & Jaberi, N. (2005). "Neuro-
    psychiatric Effects of Caffeine." *Advances in Psychiatric Treat-
    ment* 11 (6): 432–439. https://doi.org/10.1192/apt.11.6.432.

19. Brewer, Judson A. (2021). *Unwinding Anxiety: New Science
    Shows How to Break the Cycles of Worry and Fear to Heal Your
    Mind* (New York: Avery), 109.

20. Lewis, J. G. (2013). "Alcohol, Sleep, and Why You Might
    Re-think That Nightcap." *Scitable* (blog), Nature Education,
    October 28. https://www.nature.com/scitable/blog/mind-read
    /alcohol_sleep_and_why_you.

21. Griswold, M. G., Fullman, N., Hawley, C., Arian, N., Zimsen,
    S. R. M., Tymeson, H. D., Venkateswaran, V., et al. (2018).
    "Alcohol Use and Burden for 195 Countries and Territories,
    1990–2016: A Systematic Analysis for the Global Burden of
    Disease Study 2016." *Lancet* 392 (10152): 1015–1035. https://
    doi.org/10.1016/s0140-6736(18)31310-2.

22. Georgetown Behavioral Hospital. (2021). "GABA and Alcohol: How Drinking Leads to Anxiety." *Behavioral Health News* (blog), May 6. www.gbhoh.com/gaba-and-alcohol-how-drinking-leads -to-anxiety/.

23. Camden and Islington NHS Foundation Trust. "The Unhealthy Mix between Alcohol and Mental Health." Accessed October 13, 2021. www.candi.nhs.uk/news/unhealthy-mix -between-alcohol-and-mental-health.

24. Aucoin, M., & Bhardwaj, S. (2016). "Generalized Anxiety Disorder and Hypoglycemia Symptoms Improved with Diet Modification." *Case Reports in Psychiatry*, https://doi.org /10.1155/2016/7165425.

25. Straub, R. H., & Cutolo, M. (2018). "Psychoneuroimmunology— Developments in Stress Research." *Wiener Medizinische Wochenschrift* 168: 76–84. https://doi.org/10.1007/s10354-017-0574-2.

26. Environmental Working Group. (2021). "Clean Fifteen™: EWG's 2021 Shopper's Guide to Pesticides in Produce." www .ewg.org/foodnews/clean-fifteen.php.

27. University of Rochester Medical Center. (2021). "Nutrition Facts: Chicken Liver." Health Encyclopedia. https://www .urmc.rochester.edu/encyclopedia/content.aspx?-contenttypeid =76&contentid=05028-1.

28. Hunt, Janet R. (2003). "Bioavailability of Iron, Zinc, and Other Trace Minerals from Vegetarian Diets." *American Journal of Clinical Nutrition* 78 (3): 633S–639S. https://doi.org /10.1093/ajcn/78.3.633s.

29. Johnston, B. C., Zeraatkar, D., Han, M. A., Vernooij, R. W. M., Valli, C., El Dib, R., Marshall, C., et al. (2019). "Unprocessed Red Meat and Processed Meat Consumption: Dietary Guideline Recommendations from the Nutritional Recommendations (NutriRECS) Consortium." *Annals of Internal Medicine* 171 (10): 756–764. https://doi.org/10.7326/m19-1621.

30. Masters, R. C., Liese, A. D., Haffner, S. M., Wagenknecht, L. E., & Hanley, A. J. (2010). "Whole and Refined Grain Intakes Are Related to Inflammatory Protein Concentrations in Human Plasma." *Journal of Nutrition* 140 (3): 587–594. https:// doi.org/10.3945/jn.109.116640.

31. Giugliano, D., Ceriello, A., & Esposito, K. (2006). "The Effects of Diet on Inflammation: Emphasis on the Metabolic

Syndrome." *Journal of the American College of Cardiology* 48 (4): 677–685. https://doi.org/10.1016/j.jacc.2006.03.052.

32. Gross, L. S., Li, L., Ford, E. S., & Liu, S. (2004). "Increased Consumption of Refined Carbohydrates and the Epidemic of Type 2 Diabetes in the United States: An Ecologic Assessment." *American Journal of Clinical Nutrition* 79 (5): 774–779. https://doi.org/10.1093/ajcn/79.5.774.

33. Saris, W. H. M., & Foster, G. D. (2006). "Simple Carbohydrates and Obesity: Fact, Fiction and Future." *International Journal of Obesity* 30 (S3): S1–S3. https://doi.org/10.1038/sj.ijo.0803522.

34. Gentreau, M., Chuy, V., Féart, C., Samieri, C., Ritchie, K., Raymond, M., Berticat, C., & Artero, S. (2020). "Refined Carbohydrate-Rich Diet Is Associated with Long-Term Risk of Dementia and Alzheimer's Disease in Apolipoprotein E ε4 Allele Carriers." *Alzheimer's & Dementia* 16 (7): 1043–1053. https://doi.org/10.1002/alz.12114.

35. Temple, N. (2018). "Fat, Sugar, Whole Grains and Heart Disease: 50 Years of Confusion." *Nutrients* 10 (1): 39. https://doi.org/10.3390/nu10010039.

36. Swaminathan et al., "Associations of Cereal Grains Intake."

37. Marlett, J. A., McBurney, M. I., & Slavin, J. L. (2002). "Position of the American Dietetic Association: Health Implications of Dietary Fiber." *Journal of the American Dietetic Association* 102 (7): 993–1000. https://pubmed.ncbi.nlm.nih.gov/12146567/.

38. Swaminathan et al., "Associations of Cereal Grains Intake."

39. Sadeghi, O., Hassanzadeh-Keshteli, A., Afshar, H., Esmaillzadeh, A., & Adibi, P. (2017). "The Association of Whole and Refined Grains Consumption with Psychological Disorders among Iranian Adults." *European Journal of Nutrition* 58 (1): 211–225. https://doi.org/10.1007/s00394-017-1585-x.

40. Clarke, G., Fitzgerald, P., Hennessy, A. A., Cassidy, E. M., Quigley, E. M. M., Ross, P., Stanton, C., et al. (2010). "Marked Elevations in Pro-Inflammatory Polyunsaturated Fatty Acid Metabolites in Females with Irritable Bowel Syndrome." *Journal of Lipid Research* 51 (5): 1186–1192. https://doi.org/10.1194/jlr.p000695.

41. Patterson, E., Wall, R., Fitzgerald, G. F., Ross, R. P., & Stanton, C. (2012). "Health Implications of High Dietary

Omega-6 Polyunsaturated Fatty Acids." *Journal of Nutrition and Metabolism*: 1–16. https://doi.org/10.1155/2012/539426.

42. Ginter, E., & Simko, V. (2016). "New Data on Harmful Effects of Trans-Fatty Acids." *Bratislavske Lekarske Listy* 117 (5): 251–253. https://doi.org/10.4149/bll_2016_048.

43. Mozaffarian, D., Aro, A., & Willett, W. C. (2009). "Health Effects of Trans-Fatty Acids: Experimental and Observational Evidence." Supplement 2, *European Journal of Clinical Nutrition* 63: S5–S21. https://doi.org/10.1038/sj.ejcn.1602973.

44. Mozaffarian, D., Katan, M. B., Ascherio, A., Stampfer, M. J., & Willett, W. C. (2006). "Trans Fatty Acids and Cardiovascular Disease." *New England Journal of Medicine* 354 (15): 1601–1613. https://doi.org/10.1056/NEJMra054035.

45. Perumalla Venkata, R., & Subramanyam, R. (2016). "Evaluation of the Deleterious Health Effects of Consumption of Repeatedly Heated Vegetable Oil." *Toxicology Reports* 3: 636–643. https://doi.org/10.1016/j.toxrep.2016.08.003.

46. Le, T. T., Huff, T. B., & Cheng, J.-X. (2009). "Coherent Anti-Stokes Raman Scattering Imaging of Lipids in Cancer Metastasis." *BMC Cancer* 9 (42). https://doi.org/10.1186/1471-2407-9-42.

47. Strandwitz, P., Kim, K. H., Terekhova, D., Liu, J. K., Sharma, A., Levering, J., McDonald, D., et al. (2019). "GABA-Modulating Bacteria of the Human Gut Microbiota." *Nature Microbiology* 4 (3): 396–403. https://doi.org/10.1038/s41564-018-0307-3.

48. Stasi, C., Sadalla, S., & Milani, S. (2019). "The Relationship between the Serotonin Metabolism, Gut-Microbiota and the Gut-Brain Axis." *Current Drug Metabolism* 20 (8): 646–655. https://doi.org/10.2174/1389200220666190725115503.

49. Yano, J. M., Yu, K., Donaldson, G. P., Shastri, G. G., Ann, P., Ma, L., Nagler, C. R., et al. (2015). "Indigenous Bacteria from the Gut Microbiota Regulate Host Serotonin Biosynthesis." *Cell* 161 (2): 264–276. https://doi.org/10.1016/j.cell.2015.02.047.

50. Kresser, Chris. (2019). "The Bountiful Benefits of Bone Broth: A Comprehensive Guide." *Chris Kresser* (blog), August 16. https://chriskresser.com/the-bountiful-benefits-of-bone-broth-a-comprehensive-guide/#Bone_Broth_in_Traditional_Cultures.

51. Todorov, A., Chumpalova-Tumbeva, P., Stoimenova-Popova,

M., Popova, V. S., Todorieva-Todorova, D., Tzvetkov, N., Hristov, I. G., et al. (2018). "Correlation between Depression and Anxiety and the Level of Vitamin B_{12} in Patients with Depression and Anxiety and Healthy Controls." *Journal of Biomedical and Clinical Research* 10 (2): 140–145. https://doi.org/10.1515/jbcr-2017-0023.

52. Pandey, A., Dabhade, P., & Kumarasamy, A. (2019). "Inflammatory Effects of Subacute Exposure of Roundup in Rat Liver and Adipose Tissue." *Dose-Response* 17 (2). https://doi.org/10.1177/1559325819843380.

53. Vasiluk, L., Pinto, L. J., & Moore, M. M. (2005). "Oral Bioavailability of Glyphosate: Studies Using Two Intestinal Cell Lines." *Environmental Toxicology and Chemistry* 24 (1): 153. https://doi.org/10.1897/04-088r.1.

54. International Agency for Research on Cancer. (2015). "IARC Monograph on Glyphosate." www.iarc.who.int/featured-news/media-centre-iarc-news-glyphosate/.

55. Palmnäs, M. S. A., Cowan, T. E., Bomhof, M. R., Su, J., Reimer, R. A., Vogel, H. J., Hittel, D. S., & Shearer, J. (2014). "Low-Dose Aspartame Consumption Differentially Affects Gut Microbiota-Host Metabolic Interactions in the Diet-Induced Obese Rat." *PLoS ONE* 9 (10). https://doi.org/10.1371/journal.pone.0109841.

56. Gul, S. S., Hamilton, A. R. L., Munoz, A. R., Phupitakphol, T., Liu, W., Hyoju, S. J., Economopoulos, K. P., et al. (2017). "Inhibition of the Gut Enzyme Intestinal Alkaline Phosphatase May Explain How Aspartame Promotes Glucose Intolerance and Obesity in Mice." *Applied Physiology, Nutrition, and Metabolism* 42 (1): 77–83. https://doi.org/10.1139/apnm-2016-0346.

57. Claesson, A.-L., Holm, G., Ernersson, A., Lindström, T., & Nystrom, F. H. (2009). "Two Weeks of Overfeeding with Candy, but Not Peanuts, Increases Insulin Levels and Body Weight." *Scandinavian Journal of Clinical and Laboratory Investigation* 69 (5): 598–605. https://doi.org/10.1080/00365510902912754.

Chapter 8: Body on Fire

1. Amodeo, G., Trusso, M. A., & Fagiolini, A. (2018). "Depression and Inflammation: Disentangling a Clear Yet Complex

and Multifaceted Link." *Neuropsychiatry* 7 (4). https://doi .org/10.4172/neuropsychiatry.1000236.

2. Felger, J. C. (2018). "Imaging the Role of Inflammation in Mood and Anxiety-Related Disorders." *Current Neuropharmacology* 16 (5): 533–558. https://doi.org/10.2174/1570159X15 666171123201142.

3. Schiepers, O. J., Wichers, M. C., & Maes, M. (2005). "Cytokines and Major Depression." *Progress in Neuro-Psychopharmacology & Biological Psychiatry* 29 (2): 201–217. https://doi.org/10.1016/j.pnpbp.2004.11.003.

4. Felger, "Imaging the Role of Inflammation."

5. Attwells, S., Setiawan, E., Wilson, A. A., Rusjan, P. M., Mizrahi, R., Miler, L., Xu, C., et al. (2017). "Inflammation in the Neurocircuitry of Obsessive-Compulsive Disorder." *JAMA Psychiatry* 74 (8): 833–840. https://doi.org/10.1001/jamapsy chiatry.2017.1567.

6. Gerentes, M., Pelissolo, A., Rajagopal, K., Tamouza, R., & Hamdani, N. (2019). "Obsessive-Compulsive Disorder: Autoimmunity and Neuroinflammation." *Current Psychiatry Reports* 21 (8): 78. https://doi.org/10.1007/s11920-019-1062-8.

7. Johns Hopkins Medicine: Pathology. (2021). "Prevalence of Autoimmune Diseases—Autoimmune Disease." https:// pathology.jhu.edu/autoimmune/prevalence.

8. National Institutes of Health. (2021). "Autoimmunity May Be Rising in the United States." April 8. www.nih.gov/news -events/news-releases/autoimmunity-may-be-rising-united -states.

9. National Institutes of Health, "Autoimmunity May Be Rising."

10. Fasano, A. (2011). "Zonulin and Its Regulation of Intestinal Barrier Function: The Biological Door to Inflammation, Autoimmunity, and Cancer." *Physiological Reviews* 91 (1): 151–175. https://doi.org/10.1152/physrev.00003.2008.

11. Rowley, B., & Monestier, M. (2005). "Mechanisms of Heavy Metal-Induced Autoimmunity." *Molecular Immunology* 42 (7): 833–838. https://doi.org/10.1016/j.molimm .2004.07.050.

12. Harding, C., Pytte, C., Page, K., Ryberg, K., Normand, E., Remigio, G., DeStefano, R. A., et al. (2020). "Mold Inhalation Causes Innate Immune Activation, Neural, Cognitive

and Emotional Dysfunction." *Brain, Behavior, and Immunity* 87: 218–228. https://doi.org/10.1016/j.bbi.2019.11.006.

13. Benros, M. E., Waltoft, B. L., Nordentoft, M., Østergaard, S. D., Eaton, W. W., Krogh, J., & Mortensen, P. B. (2013). "Autoimmune Diseases and Severe Infections as Risk Factors for Mood Disorders: A Nationwide Study." *JAMA Psychiatry* 70 (8): 812–820. https://doi.org/10.1001/jamapsychiatry.2013.1111.

14. Dube, S. R., Fairweather, D., Pearson, W. S., Felitti, V. J., Anda, R. F., & Croft, J. B. (2009). "Cumulative Childhood Stress and Autoimmune Diseases in Adults." *Psychosomatic Medicine* 71 (2): 243–250. https://doi.org/10.1097/PSY.0b013e3181907888.

15. Vighi, G., Marcucci, F., Sensi, L., Di Cara, G., & Frati, F. (2008). "Allergy and the Gastrointestinal System." *Clinical & Experimental Immunology* 153 (S1): 3–6. https://doi.org/10.1111/j.1365-2249.2008.03713.x.

16. Bonaz, B., Bazin, T., & Pellissier, S. (2018). "The Vagus Nerve at the Interface of the Microbiota-Gut-Brain Axis." *Frontiers in Neuroscience* 12. https://doi.org/10.3389/fnins.2018.00049.

17. Petra, A. I., Panagiotidou, S., Hatziagelaki, E., Stewart, J. M., Conti, P., & Theoharides, T. C. (2015). "Gut-Microbiota-Brain Axis and Its Effect on Neuropsychiatric Disorders with Suspected Immune Dysregulation." *Clinical Therapeutics* 37 (5): 984–995. https://doi.org/10.1016/j.clinthera.2015.04.002.

18. Marin, I., Goertz, J., Ren, T., Rich, S., Onengut-Gumuscu, S., Farber, E., Wu, M., et al. (2017). "Microbiota Alteration Is Associated with the Development of Stress-Induced Despair Behavior." *Scientific Reports* 7 (1): 43859. https://doi.org/10.1038/srep43859.

19. Lurie, I., Yang, Y.-X., Haynes, K., Mamtani, R., & Boursi, B. (2015). "Antibiotic Exposure and the Risk for Depression, Anxiety, or Psychosis: A Nested Case-Control Study." *Journal of Clinical Psychiatry* 76 (11): 1522–1528. https://doi.org/10.4088/JCP.15m09961.

20. Marotta, A., Sarno, E., Del Casale, A., Pane, M., Mogna, L., Amoruso, A., Felis, G. E., & Fiorio, M. (2019). "Effects of Probiotics on Cognitive Reactivity, Mood, and Sleep Quality." *Frontiers in Psychiatry* 10: 164. https://doi.org/10.3389/fpsyt.2019.00164.

21. Kato-Kataoka, A., Nishida, K., Takada, M., Suda, K., Kawai, M., Shimizu, K., Kushiro, A., et al. (2016). "Fermented Milk Containing *Lactobacillus casei* Strain Shirota Prevents the Onset of Physical Symptoms in Medical Students under Academic Examination Stress." *Beneficial Microbes* 7 (2): 153–156. https://doi.org/10.3920/BM2015.0100.

22. Guo, Y., Xie, J.-P., Deng, K., Li, X., Yuan, Y., Xuan, Q., Xie, J., et al. (2019). "Prophylactic Effects of *Bifidobacterium adolescentis* on Anxiety and Depression-Like Phenotypes after Chronic Stress: A Role of the Gut Microbiota-Inflammation Axis." *Frontiers in Behavioral Neuroscience* 13: 126. https://doi.org/10.3389/fnbeh.2019.00126.

23. Noonan, S., Zaveri, M., Macaninch, E., & Martyn, K. (2020). "Food & Mood: A Review of Supplementary Prebiotic and Probiotic Interventions in the Treatment of Anxiety and Depression in Adults." *BMJ Nutrition, Prevention & Health* 3 (2): 351–362. https://doi.org/10.1136/bmjnph-2019-000053.

24. Strandwitz, P., Kim, K. H., Terekhova, D., Liu, J. K., Sharma, A., Levering, J., McDonald D., et al. (2018). "GABA-Modulating Bacteria of the Human Gut Microbiota." *Nature Microbiology* 4 (3): 396–403. https://doi.org/10.1038/s41564-018-0307-3.

25. Guo et al., "Prophylactic Effects of *Bifidobacterium adolescentis* on Anxiety."

26. Daulatzai, M. (2015). "Non-Celiac Gluten Sensitivity Triggers Gut Dysbiosis, Neuroinflammation, Gut-Brain Axis Dysfunction, and Vulnerability for Dementia." *CNS & Neurological Disorders—Drug Targets* 14 (1): 110–131. www.ingentaconnect.com/content/ben/cnsnddt/2015/00000014/00000001/art00018#Refs.

27. Kaliannan, K., Wang, B., Li, X.-Y., Kim, K.-J., & Kang, J. X. (2015). "A Host-Microbiome Interaction Mediates the Opposing Effects of Omega-6 and Omega-3 Fatty Acids on Metabolic Endotoxemia." *Scientific Reports* 5. https://doi.org/10.1038/srep11276.

28. Scaioli, E., Liverani, E., & Belluzzi, A. (2017). "The Imbalance between N-6/N-3 Polyunsaturated Fatty Acids and Inflammatory Bowel Disease: A Comprehensive Review and Future

Therapeutic Perspectives." *International Journal of Molecular Sciences* 18 (12): 2619. https://doi.org/10.3390/ijms18122619.

29. Clarke, G., Fitzgerald, P., Hennessy, A. A., Cassidy, E. M., Quigley, E. M. M., Ross, P., Stanton, C., et al. (2010). "Marked Elevations in Pro-Inflammatory Polyunsaturated Fatty Acid Metabolites in Females with Irritable Bowel Syndrome." *Journal of Lipid Research* 51 (5): 1186–1192. https://doi.org/10.1194/jlr.P000695.

30. Shil, A., & Chichger, H. (2021). "Artificial Sweeteners Negatively Regulate Pathogenic Characteristics of Two Model Gut Bacteria, *E. coli* and *E. faecalis*." *International Journal of Molecular Sciences* 22 (10): 5228. https://doi.org/10.3390/ijms22105228.

31. Wu, W., Zhou, J., Chen, J., Han, H., Liu, J., Niu, T., & Weng, F. (2020). "Dietary κ-Carrageenan Facilitates Gut Microbiota-Mediated Intestinal Inflammation." Preprint, submitted August 18. https://doi.org/10.21203/rs.3.rs-56671/v1.

32. Aitbali, Y., Ba-M'hamed, S., Elhidar, N., Nafis, A., Soraa, N., & Bennis, M. (2018). "Glyphosate-Based Herbicide Exposure Affects Gut Microbiota, Anxiety and Depression-Like Behaviors in Mice." *Neurotoxicology and Teratology* 67: 44–49. https://doi.org/10.1016/j.ntt.2018.04.002.

33. Imhann, F., Bonder, M. J., Vich Vila, A., Fu, J., Mujagic, Z., Vork, L., Tigchelaar, E. F., et al. (2016). "Proton Pump Inhibitors Affect the Gut Microbiome." *Gut* 65 (5): 740–748. https://doi.org/10.1136/gutjnl-2015-310376.

34. Rogers, M. A. M., & Aronoff, D. M. (2015). "The Influence of Non-Steroidal Anti-Inflammatory Drugs on the Gut Microbiome." *Clinical Microbiology and Infection* 22 (2): 178. e1–178.e9. https://doi.org/10.1016/j.cmi.2015.10.003.

35. Camilleri, M., Lembo, A., & Katzka, D. A. (2017). "Opioids in Gastroenterology: Treating Adverse Effects and Creating Therapeutic Benefits." *Clinical Gastroenterology and Hepatology* 15 (9): 1338–1349. https://doi.org/10.1016/j.cgh.2017.05.014.

36. Khalili, H. (2015). "Risk of Inflammatory Bowel Disease with Oral Contraceptives and Menopausal Hormone Therapy: Current Evidence and Future Directions." *Drug Safety* 39 (3): 193–197. https://doi.org/10.1007/s40264-015-0372-y.

37. Levy, J. (2000). "The Effects of Antibiotic Use on Gastro-intestinal Function." *American Journal of Gastroenterology* 95 (1 Suppl.): S8–S10. https://doi.org/10.1016/s0002-9270(99)00808-4.

38. Olivera, A., Moore, T. W., Hu, F., Brown, A. P., Sun, A., Liotta, D. C., Snyder, J. P., et al. (2012). "Inhibition of the NF-κB Signaling Pathway by the Curcumin Analog, 3,5-Bis(2-Pyridinylmethylidene)-4-piperidone (EF31): Anti-Inflammatory and Anti-Cancer Properties." *International Immunopharmacology* 12 (2): 368–377. https://doi.org/10.1016/j.intimp.2011.12.009.

39. Chainani-Wu, Nita. (2003). "Safety and Anti-Inflammatory Activity of Curcumin: A Component of Tumeric (*Curcuma longa*)." *Journal of Alternative and Complementary Medicine* 9 (1): 161–168. https://doi.org/10.1089/107555303321223035.

40. Grzanna, R., Lindmark, L., & Frondoza, C. G. (2005). "Ginger—An Herbal Medicinal Product with Broad Anti-Inflammatory Actions." *Journal of Medicinal Food* 8 (2): 125–132. https://doi.org/10.1089/jmf.2005.8.125.

41. Arreola, R., Quintero-Fabián, S., López-Roa, R. I., Flores-Gutiérrez, E. O., Reyes-Grajeda, J. P., Carrera-Quintanar, L., & Ortuño-Sahagún, D. (2015). "Immunomodulation and Anti-Inflammatory Effects of Garlic Compounds." *Journal of Immunology Research* 2015: 1–13. https://doi.org/10.1155/2015/401630.

42. Dorsch, W., Schneider, E., Bayer, T., Breu, W., & Wagner, H. (1990). "Anti-Inflammatory Effects of Onions: Inhibition of Chemotaxis of Human Polymorphonuclear Leukocytes by Thiosulfinates and Cepaenes." *International Archives of Allergy and Applied Immunology* 92 (1): 39–42. https://doi.org/10.1159/000235221.

43. Calder, Philip C. (2010). "Omega-3 Fatty Acids and Inflammatory Processes." *Nutrients* 2 (3): 355–374. https://doi.org/10.3390/nu2030355.

44. Zhu, F., Du, B., & Xu, B. (2017). "Anti-Inflammatory Effects of Phytochemicals from Fruits, Vegetables, and Food Legumes: A Review." *Critical Reviews in Food Science and Nutrition* 58 (8): 1260–1270. https://doi.org/10.1080/10408398.2016.1251390.

45. Centers for Disease Control and Prevention (2021). "Births—
     Method of Delivery." FastStats, CDC. www.cdc.gov/nchs
     /fastats/delivery.htm.

46. Shin, H., Pei, Z., Martinez II, K. A., Rivera-Vinas, J. I.,
     Mendez, K., Cavallin, H., & Dominguez-Bello, M. G.
     (2015). "The First Microbial Environment of Infants Born by
     C-Section: The Operating Room Microbes." *Microbiome* 3.
     https://doi.org/10.1186/s40168-015-0126-1.

47. Ledger, W. J., & Blaser, M. J. (2013). "Are We Using Too
     Many Antibiotics during Pregnancy?" *BJOG: An International
     Journal of Obstetrics and Gynaecology* 120 (12): 1450–1452.
     https://doi.org/10.1111/1471-0528.12371.

48. Blaser, Martin J. (2014). *Missing Microbes: How the Overuse of
     Antibiotics Is Fueling Our Modern Plagues* (New York: Henry
     Holt), 219.

49. Prescott, J. (2015). "[Review of] *Missing Microbes: How the
     Overuse of Antibiotics Is Fueling Our Modern Plagues*." *Cana-
     dian Veterinary Journal* 56 (12): 1260.

50. Anand, D., Colpo, G. D., Zeni, G., Zeni, C. P., & Teixeira,
     A. L. (2017). "Attention-Deficit/Hyperactivity Disorder and
     Inflammation: What Does Current Knowledge Tell Us? A
     Systematic Review." *Frontiers in Psychiatry* 8: 228. https://doi
     .org/10.3389/fpsyt.2017.00228.

51. Yudkin, J. S., Kumari, M., Humphries, S. E., & Mohamed-
     Ali, V. (2000). "Inflammation, Obesity, Stress and Coronary
     Heart Disease: Is Interleukin-6 the Link?" *Atherosclerosis* 148
     (2): 209–214. https://doi.org/10.1016/s0021-9150(99)00463-3.

52. Grivennikov, S. I., Greten, F. R., & Karin, M. (2010). "Im-
     munity, Inflammation, and Cancer." *Cell* 140 (6): 883–899.
     https://doi.org/10.1016/j.cell.2010.01.025.

53. Leonard, B. E. (2007). "Inflammation, Depression and
     Dementia: Are They Connected?" *Neurochemical Research*
     32 (10): 1749–1756. https://doi.org/10.1007/s11064-007
     -9385-y.

54. Berk, M., Williams, L. J., Jacka, F. N., O'Neil, A., Pasco, J. A.,
     Moylan, S., Allen, N. B., et al. (2013). "So Depression Is an
     Inflammatory Disease, but Where Does the Inflammation
     Come From?" *BMC Medicine* 11 (1): 200. https://doi.org/10.1186
     /1741-7015-11-200.

55. Felger, "Imaging the Role of Inflammation in Mood and Anxiety-Related Disorders."

56. Jolliffe, D. A., Camargo, C. A., Sluyter, J. D., Aglipay, M., Aloia, J. F., Ganmaa, D., Bergman P., et al. (2021). "Vitamin D Supplementation to Prevent Acute Respiratory Infections: A Systematic Review and Meta-Analysis of Aggregate Data from Randomised Controlled Trials." *The Lancet Diabetes & Endocrinology* 9 (5): 276–292. https://doi.org/10.1016/S2213-8587(21)00051-6.

57. Picotto, G., Liaudat, A. C., Bohl, L., & Tolosa de Talamoni, N. (2012). "Molecular Aspects of Vitamin D Anticancer Activity." *Cancer Investigation* 30 (8): 604–614. https://doi.org/10.3109/07357907.2012.721039.

58. Martineau, A. R., Jolliffe, D. A., Hooper, R. L., Greenberg, L., Aloia, J. F., Bergman, P., Dubnov-Raz, G., et al. (2017). "Vitamin D Supplementation to Prevent Acute Respiratory Tract Infections: Systematic Review and Meta-Analysis of Individual Participant Data." BMJ 2017 (356): i6583. https://doi.org/10.1136/bmj.i6583.

59. Akbar, N. A., & Zacharek, M. A. (2011). "Vitamin D: Immunomodulation of Asthma, Allergic Rhinitis, and Chronic Rhinosinusitis." *Current Opinion in Otolaryngology and Head and Neck Surgery* 19 (3): 224–228. https://doi.org/10.1097/MOO.0b013e3283465687.

60. Aranow, C. (2011). "Vitamin D and the Immune System." *Journal of Investigative Medicine: The Official Publication of the American Federation for Clinical Research* 59 (6): 881–886. https://doi.org/10.2310/JIM.0b013e31821b8755.

61. Littlejohns, T. J., Henley, W. E., Lang, I. A., Annweiler, C., Beauchet, O., Chaves, P. H. M., Fried, L., et al. (2014). "Vitamin D and the Risk of Dementia and Alzheimer Disease." *Neurology* 83 (10): 920–928. https://doi.org/10.1212/wnl.0000000000000755.

62. Wang, T. J., Pencina, M. J., Booth, S. L., Jacques, P. F., Ingelsson, E., Lanier, K., Benjamin, E. J., et al. (2008). "Vitamin D Deficiency and Risk of Cardiovascular Disease." *Circulation* 117 (4): 503–511. https://doi.org/10.1161/circulationaha.107.706127.

63. Lips, P., & van Schoor, N. M. (2011). "The Effect of Vitamin D on Bone and Osteoporosis." *Best Practice & Research Clinical Endocrinology & Metabolism* 25 (4): 585–591. https://doi.org/10.1016/j.beem.2011.05.002.

64. Pilz, S., Zittermann, A., Obeid, R., Hahn, A., Pludowski, P., Trummer, C., Lerchbaum, E., et al. (2018). "The Role of Vitamin D in Fertility and during Pregnancy and Lactation: A Review of Clinical Data." *International Journal of Environmental Research and Public Health* 15 (10): 2241. https://doi.org/10.3390/ijerph15102241.

65. Picotto et al., "Molecular Aspects of Vitamin D Anticancer Activity."

66. Garland, C. F., Garland, F. C., Gorham, E. D., Lipkin, M., Newmark, H., Mohr, S. B., & Holick, M. F. (2006). "The Role of Vitamin D in Cancer Prevention." *American Journal of Public Health* 96 (2): 252–261. https://doi.org/10.2105/ajph.2004.045260.

67. Fleet, J. C., DeSmet, M., Johnson, R., & Li, Y. (2012). "Vitamin D and Cancer: A Review of Molecular Mechanisms." *Biochemical Journal* 441 (1): 61–76. https://doi.org/10.1042/BJ20110744.

68. Hargrove, L., Francis, T., & Francis, H. (2014), "Vitamin D and GI Cancers: Shedding Some Light on Dark Diseases." *Annals of Translational Medicine* 2 (1): 9. https://doi.org/10.3978/j.issn.2305-5839.2013.03.04.

69. Vuolo, L., Di Somma, C., Faggiano, A., & Colao, A. (2012). "Vitamin D and Cancer." *Frontiers in Endocrinology* 3: 58. https://doi.org/10.3389/fendo.2012.00058.

70. Chakraborti, C. K. (2011). "Vitamin D as a Promising Anticancer Agent." *Indian Journal of Pharmacology* 43 (2): 113–120. https://doi.org/10.4103/0253-7613.77335.

71. Menon, V., Kar, S. K., Suthar, N., & Nebhinani, N. (2020). "Vitamin D and Depression: A Critical Appraisal of the Evidence and Future Directions." *Indian Journal of Psychological Medicine* 42 (1): 11–21. https://doi.org/10.4103/IJPSYM.IJPSYM_160_19.

72. Armstrong, D. J., Meenagh, G. K., Bickle, I., Lee, A. S. H., Curran, E.-S., & Finch, M. B. (2007). "Vitamin D Deficiency Is Associated with Anxiety and Depression in Fibro-

myalgia." *Clinical Rheumatology* 26 (4): 551–554. https://doi
.org/10.1007/s10067-006-0348-5.

73. Parva, N. R., Tadepalli, S., Singh, P., Qian, A., Joshi, R.,
Kandala, H., Nookala, V. K., & Cheriyath, P. (2018). "Prevalence of Vitamin D Deficiency and Associated Risk Factors
in the US Population (2011–2012)." *Cureus* 10 (6). https://doi
.org/10.7759/cureus.2741.

74. Mithal, A., Wahl, D. A., Bonjour, J.-P., Burckhardt, P.,
Dawson-Hughes, B., Eisman, J. A., El-Hajj Fuleihan, G., et
al. (2009). "Global Vitamin D Status and Determinants of
Hypovitaminosis D." *Osteoporosis International* 20 (11): 1807–
1820. https://doi.org/10.1007/s00198-009-0954-6.

75. Kumar, J., Muntner, P., Kaskel, F. J., Hailpern, S. M., &
Melamed, M. L. (2009). "Prevalence and Associations of
25-Hydroxyvitamin D Deficiency in US Children: NHANES
2001–2004." *Pediatrics* 124 (3): e362–e370. https://doi.org
/10.1542/peds.2009-0051.

76. Amrein, K., Scherkl, M., Hoffmann, M., Neuwersch-
Sommeregger, S., Köstenberger, M., Tmava Berisha, A., Mar-
tucci, G., et al. (2020). "Vitamin D Deficiency 2.0: An Update
on the Current Status Worldwide." *European Journal of Clinical
Nutrition* 74 (11): 1498–1513. https://doi.org/10.1038/s41430
-020-0558-y.

77. Bradford, P. T. (2009). "Skin Cancer in Skin of Color." *Dermatology Nursing* 21 (4): 170–178. https://www.ncbi.nlm.nih
.gov/pmc/articles/PMC2757062.

78. University of Pennsylvania. (2017). "Genes Responsible for
Diversity of Human Skin Colors Identified." *ScienceDaily*,
October 12. www.sciencedaily.com/releases/2017/10/1710121
43324.htm.

79. University of Pennsylvania, "Genes Responsible."

80. Bradford, "Skin Cancer in Skin of Color."

81. Brenner, M., & Hearing, V. J. (2008). "The Protective Role of
Melanin against UV Damage in Human Skin." *Photochemis-
try and Photobiology* 84 (3): 539–549. https://doi.org/10.1111
/j.1751-1097.2007.00226.x.

82. Montagna, W., & Carlisle, K. (1991). "The Architecture of
Black and White Facial Skin." *Journal of the American Academy*

of Dermatology 24 (6): 929–937. https://doi.org/10.1016/0190
-9622(91)70148-u.

83. Mezza, T., Muscogiuri, G., Sorice, G. P., Prioletta, A., Salo-
mone, E., Pontecorvi, A., & Giaccari, A. (2012). "Vitamin
D Deficiency: A New Risk Factor for Type 2 Diabetes?" *Annals of Nutrition & Metabolism* 61 (4): 337–348. https://doi
.org/10.1159/000342771.

84. Martin, T., & Campbell, R. K. (2011). "Vitamin D and
Diabetes." *Diabetes Spectrum* 24 (2): 113–118. https://doi
.org/10.2337/diaspect.24.2.113.

85. Marks, R. (2020). "Obesity, COVID-19 and Vitamin D: Is
There an Association Worth Examining?" *Advances in Obesity, Weight Management & Control* 10 (3): 59–63. https://doi.
org/10.15406/aowmc.2020.10.00307.

86. Castillo, M. E., Costa, L. M. E., Barrios, J. M. V., Díaz,
J. F. A., Miranda, J. L., Bouillon, R., & Gomez, J. M. Q.
(2020). "Effect of Calcifediol Treatment and Best Available
Therapy versus Best Available Therapy on Intensive Care
Unit Admission and Mortality among Patients Hospitalized
for COVID-19: A Pilot Randomized Clinical Study." *Journal
of Steroid Biochemistry and Molecular Biology* 203. https://doi
.org/10.1016/j.jsbmb.2020.105751.

87. Meltzer, D. O., Best, T. J., Zhang, H., Vokes, T., Arora, V.,
& Solway, J. (2020). "Association of Vitamin D Status and
Other Clinical Characteristics with COVID-19 Test Results." *JAMA Network Open* 3 (9). https://doi.org/10.1001
/jamanetworkopen.2020.19722.

88. Littlejohns et al., "Vitamin D and the Risk of Dementia and
Alzheimer Disease."

89. Garland et al., "The Role of Vitamin D in Cancer Prevention."

90. Bilinski, K., & Boyages, J. (2013). "Association between
25-Hydroxyvitamin D Concentration and Breast Cancer Risk
in an Australian Population: An Observational Case-Control
Study." *Breast Cancer Research and Treatment* 137 (2): 599–
607. https://doi.org/10.1007/s10549-012-2381-1.

91. Holick, M. F. (2004). "Sunlight and Vitamin D for Bone
Health and Prevention of Autoimmune Diseases, Cancers,
and Cardiovascular Disease." Supplement, *American Journal of*

*ClinicalNutrition*80(6):1678S–1688S.https://doi.org/10.1093/ajcn/80.6.1678S.

92. Brøndum-Jacobsen, P., Benn, M., Jensen, G. B., & Nordestgaard, B. G. (2012). "25-Hydroxyvitamin D Levels and Risk of Ischemic Heart Disease, Myocardial Infarction, and Early Death: Population-Based Study and Meta-Analyses of 18 and 17 Studies." *Arteriosclerosis, Thrombosis, and Vascular Biology* 32 (11): 2794–2802. https://doi.org/10.1161/ATV BAHA.112.248039.

93. Wang et al., "Vitamin D Deficiency and Risk of Cardiovascular Disease."

94. Lips & van Schoor, "The Effect of Vitamin D on Bone and Osteoporosis."

95. Brehm, J. M., Celedón, J. C., Soto-Quiros, M. E., Avila, L., Hunninghake, G. M., Forno, E., Laskey, D., et al. (2009). "Serum Vitamin D Levels and Markers of Severity of Childhood Asthma in Costa Rica." *American Journal of Respiratory and Critical Care Medicine* 179 (9): 765–771. https://doi.org/10.1164/rccm.200808-1361OC.

96. Munger, K. L., Levin, L. I., Hollis, B. W., Howard, N. S., & Ascherio, A. (2006). "Serum 25-Hydroxyvitamin D Levels and Risk of Multiple Sclerosis." *JAMA* 296 (23): 2832–2838. https://doi.org/10.1001/jama.296.23.2832.

97. Kriegel, M. A., Manson, J. E., & Costenbader, K. H. (2011). "Does Vitamin D Affect Risk of Developing Autoimmune Disease?: A Systematic Review." *Seminars in Arthritis and Rheumatism* 40 (6): 512–531. https://doi.org/10.1016/j.semarthrit.2010.07.009.

98. Anglin, R. E. S., Samaan, Z., Walter, S. D., & McDonald, S. D. (2013). "Vitamin D Deficiency and Depression in Adults: Systematic Review and Meta-Analysis." *British Journal of Psychiatry* 202 (2): 100–107. https://doi.org/10.1192/bjp.bp.111.106666.

99. Armstrong et al., "Vitamin D Deficiency Is Associated with Anxiety and Depression in Fibromyalgia."

100. Hansen, J. P., Pareek, M., Hvolby, A., Schmedes, A., Toft, T., Dahl, E., & Nielsen, C. T. (2019). "Vitamin D3 Supplementation and Treatment Outcomes in Patients with Depression (D3-Vit-Dep)." *BMC Research Notes* 12 (1): 203. https://doi.org/10.1186/s13104-019-4218-z.

101. Lansdowne, A. T. G., & Provost, S. C. (1998). "Vitamin D3 Enhances Mood in Healthy Subjects during Winter." *Psychopharmacology* 135 (4): 319–323. https://doi.org/10.1007/s002130050517.

102. Mead, M. N. (2008). "Benefits of Sunlight: A Bright Spot for Human Health." *Environmental Health Perspectives* 116 (4): A160–A167. https://doi.org/10.1289/ehp.116-a160.

103. Kresser, C. (2021). "Vitamin D: More Is Not Better." *Chris Kresser* (blog), June 12. https://chriskresser.com/vitamin-d -more-is-not-better/.

104. Sprouse-Blum, A. S., Smith, G., Sugai, D., & Parsa, F. D. (2010). "Understanding Endorphins and Their Importance in Pain Management." *Hawaii Medical Journal* 69 (3): 70–71. https://www.ncbi.nlm.nih.gov/pmc/articles/PMC3104618.

105. Fell, G. L., Robinson, K. C., Mao, J., Woolf, C. J., & Fisher, D. E. (2014). "Skin β-Endorphin Mediates Addiction to UV Light." *Cell* 157 (7): 1527–1534. https://doi .org/10.1016/j.cell.2014.04.032.

106. Smillie, S. J., King, R., Kodji, X., Outzen, E., Pozsgai, G., Fernandes, E., Marshall, N., et al. (2014). "An Ongoing Role of α-Calcitonin Gene-Related Peptide as Part of a Protective Network against Hypertension, Vascular Hypertrophy, and Oxidative Stress." *Hypertension* 63 (5): 1056–1062. https:// doi.org/10.1161/HYPERTENSIONAHA.113.02517.

107. Staniek, V., Liebich, C., Vocks, E., Odia, S. G., Doutremepuich, J. D., Ring, J., Claudy, A., et al. (1998). "Modulation of Cutaneous SP Receptors in Atopic Dermatitis after UVA Irradiation." *Acta Dermato-Venereologica* 78 (2): 92–94. https:// doi.org/10.1080/000155598433386.

108. Pavlovic, S., Liezmann, C., Blois, S. M., Joachim, R., Kruse, J., Romani, N., Klapp, B. F., & Peters, E. M. J. (2010). "Substance P Is a Key Mediator of Stress-Induced Protection from Allergic Sensitization via Modified Antigen Presentation." *Journal of Immunology* 186 (2): 848–855. https://doi .org/10.4049/jimmunol.0903878.

109. Holliman, G., Lowe, D., Cohen, H., Felton, S., & Raj, K. (2017). "Ultraviolet Radiation-Induced Production of Nitric Oxide: A Multi-Cell and Multi-Donor Analysis." *Scientific*

Reports 7 (1): 11105. https://doi.org/10.1038/s41598-017 -11567-5.

110. Lindqvist, P. G., Epstein, E., Nielsen, K., Landin-Olsson, M., Ingvar, C., & Olsson, H. (2016). "Avoidance of Sun Exposure as a Risk Factor for Major Causes of Death: A Competing Risk Analysis of the Melanoma in Southern Sweden Cohort." *Journal of Internal Medicine* 280 (4): 375–387. https://doi .org/10.1111/joim.12496.

111. Lindqvist et al. "Avoidance of Sun Exposure as a Risk Factor."

112. Aziz, I., Lewis, N. R., Hadjivassiliou, M., Winfield, S. N., Rugg, N., Kelsall, A., Newrick, L., & Sanders, D. S. (2014). "A UK Study Assessing the Population Prevalence of Self-Reported Gluten Sensitivity and Referral Characteristics to Secondary Care." *European Journal of Gastroenterology & Hepatology* 26 (1): 33–39. https://doi.org/10.1097/01.meg.0000435546.87251.f7.

113. *Industrial Safety and Hygiene News.* (2021). "Another Country Bans Glyphosate Use." January 21. www.ishn.com/articles /112144-another-country-bans-glyphosate-use.

114. Reuters staff. (2021). "German Cabinet Approves Legislation to Ban Glyphosate from 2024." Reuters, February 10. www .reuters.com/article/us-germany-farming-lawmaking/german -cabinet-approves-legislation-to-ban-glyphosate-from-2024 -idUSKBN2AA1GF.

115. Samsel, A., & Seneff, S. (2013). "Glyphosate, Pathways to Modern Diseases II: Celiac Sprue and Gluten Intolerance." *Interdisciplinary Toxicology* 6 (4): 159–184. https://doi.org /10.2478/intox-2013-0026.

116. Center for Biological Diversity. (2020). "EPA Finds Gly-phosate Is Likely to Injure or Kill 93% of Endangered Species." November 25. https://biologicaldiversity.org/w/news/press -releases/epa-finds-glyphosate-likely-injure-or-kill-93-endangered -species-2020-11-25.

117. Wong, K. V. (2017). "Gluten and Thyroid Health." *Juniper Online Journal of Public Health* 1 (3). https://doi.org/10.19080 /jojph.2017.01.555563.

118. Benvenga, S., & Guarneri, F. (2016). "Molecular Mimicry and Autoimmune Thyroid Disease." *Reviews in Endocrine & Metabolic Disorders* 17 (4): 485–498. https://doi.org/10.1007 /s11154-016-9363-2.

119. International Agency for Research on Cancer. (2015). "IARC Monograph on Glyphosate." www.iarc.who.int/featured-news /media-centre-iarc-news-glyphosate/.

120. Caio, G., Volta, U., Tovoli, F., & De Giorgio, R. (2014). "Effect of Gluten Free Diet on Immune Response to Gliadin in Patients with Non-Celiac Gluten Sensitivity." *BMC Gastroenterology* 14 (1): 26. https://doi.org/10.1186/1471-230x-14-26.

121. Hillman, M., Weström, B., Aalaei, K., Erlanson-Albertsson, C., Wolinski, J., Lozinska, L., Sjöholm, I., et al. (2019). "Skim Milk Powder with High Content of Maillard Reaction Products Affect Weight Gain, Organ Development and Intestinal Inflammation in Early Life in Rats." *Food and Chemical Toxicology* 125: 78–84. https://doi.org/10.1016/j.fct.2018.12.015.

122. Fukudome, S., & Yoshikawa, M. (1992). "Opioid Peptides Derived from Wheat Gluten: Their Isolation and Characterization." *FEBS Letters* 296 (1): 107–111. https://doi.org /10.1016/0014-5793(92)80414-c.

123. Trivedi, M., Zhang, Y., Lopez-Toledano, M., Clarke, A., & Deth, R. (2016). "Differential Neurogenic Effects of Casein-Derived Opioid Peptides on Neuronal Stem Cells: Implications for Redox-Based Epigenetic Changes." *Journal of Nutritional Biochemistry* 37: 39–46. https://doi.org/10.1016/j .jnutbio.2015.10.012.

124. Liu, Z., & Udenigwe, C. C. (2018). "Role of Food-Derived Opioid Peptides in the Central Nervous and Gastrointestinal Systems." *Journal of Food Biochemistry* 43 (1). https://doi.org /10.1111/jfbc.12629.

125. Trivedi, M. S., Shah, J. S., Al-Mughairy, S., Hodgson, N. W., Simms, B., Trooskens, G. A., Van Criekinge, W., & Deth, R. C. (2014). "Food-Derived Opioid Peptides Inhibit Cysteine Uptake with Redox and Epigenetic Consequences." *Journal of Nutritional Biochemistry* 25 (10): 1011–1018. https://doi .org/10.1016/j.jnutbio.2014.05.004.

126. ScienceDirect. "Casomorphin." (2021). www.sciencedirect.com /topics/agricultural-and-biological-sciences/casomorphin.

127. Teschemacher, H., Koch, G., & Brantl, V. (1997). "Milk Protein-Derived Opioid Receptor Ligands." *Biopolymers* 43 (2): 99–117. https://doi.org/10.1002/(SICI)1097-0282(1997) 43:2<99::AID-BIP3>3.0.CO;2-V.

128. Goldmeier, D., Garvey, L., & Barton, S. (2008). "Does Chronic Stress Lead to Increased Rates of Recurrences of Genital Herpes—A Review of the Psychoneuroimmunological Evidence?" *International Journal of STD & AIDS* 19 (6): 359–362. https:// doi.org/10.1258/ijsa.2007.007304.

129. Mindel, A., & Marks, C. (2005). "Psychological Symptoms Associated with Genital Herpes Virus Infections: Epidemiology and Approaches to Management." *CNS Drugs* 19 (4): 303–312. https://doi.org/10.2165/00023210-200519040-00003.

Chapter 9: Women's Hormonal Health and Anxiety

1. Tasca, C., Rapetti, M., Carta, M. G., & Fadda, B. (2012). "Women and Hysteria in the History of Mental Health." *Clinical Practice and Epidemiology in Mental Health* 8: 110–19. https://dx.doi.org/10.2174%2F1745017901208010110.

2. Minerbi, A., & Fitzcharles, M. A. (2020). "Gut Microbiome: Pertinence in Fibromyalgia." Supplement 123, *Clinical and Experimental Rheumatology* 38 (1): 99–104. https://pubmed .ncbi.nlm.nih.gov/32116215/.

3. Myhill, S., Booth, N. E., & McLaren-Howard, J. (2009). "Chronic Fatigue Syndrome and Mitochondrial Dysfunction." *International Journal of Clinical and Experimental Medicine* 2 (1): 1–16. https://pubmed.ncbi.nlm.nih.gov/19436827.

4. Bartels, E. M., Dreyer, L., Jacobsen, S., Jespersen, A., Bliddal, H., & Danneskiold-Samsøe, B. (2009). "Fibromyalgi, diagnostik og praevalens. Kan kønsforskellen forklares?" [Fibromyalgia, Diagnosis and Prevalence. Are Gender Differences Explainable?]. *Ugeskr Laeger* 171 (49): 3588–3592. https:// pubmed.ncbi.nlm.nih.gov/19954696/.

5. American Thyroid Association. "General Information/Press Room." Accessed August 19, 2021. www.thyroid.org/media-main /press-room/.

6. American Thyroid Association, "General Information/Press Room."

7. Harvard Health. (2021). "The Lowdown on Thyroid Slowdown." August 17. www.health.harvard.edu/diseases-and-conditions /the-lowdown-on-thyroid-slowdown.

8. Chiovato, L., Magri, F., & Carlé, A. (2019). "Hypothyroidism in Context: Where We've Been and Where We're Going." *Ad-*

    vances in Therapy 36: 47–58. https://doi.org/10.1007/s12325
    -019-01080-8.

9.   Mayo Clinic. (2021). "Premenstrual Syndrome (PMS)—
     Symptoms and Causes." www.mayoclinic.org/diseases-conditions
     /premenstrual-syndrome/symptoms-causes/syc-20376780.

10. Dodson, R. E., Nishioka, M., Standley, L. J., Perovich, L. J.,
     Brody, J. G., & Rudel, R. A. (2012). "Endocrine Disruptors
     and Asthma-Associated Chemicals in Consumer Products."
     Environmental Health Perspectives 120 (7): 935–943. https://
     doi.org/10.1289/ehp.1104052.

11. Peinado, Francisco M., Iribarne-Durán, Luz M., Ocón-Hernández,
     Olga, Olea, Nicolás, & Artacho-Cordón, Francisco. (2020).
     "Endocrine Disrupting Chemicals in Cosmetics and Personal
     Care Products and Risk of Endometriosis." IntechOpen, Feb-
     ruary 25. https://www.intechopen.com/chapters/72654.

12. Patel, S. (2017). "Fragrance Compounds: The Wolves in
     Sheep's Clothings." *Medical Hypotheses* 102: 106–111. https://
     doi.org/10.1016/j.mehy.2017.03.025.

13. Dodson et al., "Endocrine Disruptors and Asthma-Associated
     Chemicals in Consumer Products."

14. Weatherly, L. M., & Gosse, J. A. (2017). "Triclosan Expo-
     sure, Transformation, and Human Health Effects." *Journal of
     Toxicology and Environmental Health. Part B, Critical Reviews*
     20 (8): 447–469. https://doi.org/10.1080/10937404.2017
     .1399306.

15. Rowdhwal, S. S. S., & Chen, J. (2018). "Toxic Effects of Di-2-
     ethylhexyl Phthalate: An Overview." *BioMed Research Interna-
     tional,* 1750368. https://doi.org/10.1155/2018/1750368.

16. Hormann, A. M., Vom Saal, F. S., Nagel, S. C., Stahlhut, R. W.,
     Moyer, C. L., Ellersieck, M. R., Welshons, W. V., Toutain,
     P. L., & Taylor, J. A. (2014). "Holding Thermal Receipt Pa-
     per and Eating Food after Using Hand Sanitizer Results in
     High Serum Bioactive and Urine Total Levels of Bisphenol A
     (BPA)." *PLoS ONE* 9 (10): e110509. https://doi.org/10.1371
     /journal.pone.0110509.

17. Hayes, T. B., Khoury, V., Narayan, A., Nazir, M., Park,
     A., Brown, T., Adame, L., et al. (2010). "Atrazine Induces
     Complete Feminization and Chemical Castration in Male
     African Clawed Frogs (*Xenopus laevis*)." *Proceedings of the*

National Academy of Sciences 107 (10): 4612–4617. https://doi
.org/10.1073/pnas.0909519107.

18. Sanders, R. (2010). "Pesticide Atrazine Can Turn Male Frogs
into Females." Berkeley News, March 1. https://news.berkeley
.edu/2010/03/01/frogs/.

19. Berg, J. M., Tymoczko, J. L., & Stryer, L. (2002). "Important
Derivatives of Cholesterol Include Bile Salts and Steroid Hormones," in *Biochemistry*, 5th ed. (New York: W. H. Freeman).
www.ncbi.nlm.nih.gov/books/NBK22339/.

20. Solano, M. E., & Arck, P. C. (2020). "Steroids, Pregnancy
and Fetal Development." *Frontiers in Immunology* 10. https://
doi.org/10.3389/fimmu.2019.03017.

21. Pickworth, C. K. (2016). "Women's Health and Hormonal
Axes." Women in Balance Institute. https://womeninbalance
.org/2016/12/13/womens-health-and-hormonal-axes/.

22. Skovlund, C. W., Mørch, L. S., Kessing, L. V., & Lidegaard, Ø.
(2016). "Association of Hormonal Contraception with Depression." *JAMA Psychiatry* 73 (11): 1154–1162. https://doi
.org/10.1001/jamapsychiatry.2016.2387. Erratum in *JAMA
Psychiatry* 74 (7): 764. https://doi.org/10.1001/jamapsychiatry
.2017.1446.

23. Anderl, C., Li, G., & Chen, F. S. (2020). "Oral Contraceptive
Use in Adolescence Predicts Lasting Vulnerability to Depression in Adulthood." *Journal of Child Psychology and Psychiatry*
61 (2): 148–156. https://doi.org/10.1111/jcpp.13115.

24. Williams, W. V. (2017). "Hormonal Contraception and the
Development of Autoimmunity: A Review of the Literature."
Linacre Quarterly 84 (3): 275–295. https://doi.org/10.1080/0
0243639.2017.1360065.

25. Palmery, M., Saraceno, A., Vaiarelli, A., & Carlomagno, G.
(2013). "Oral Contraceptives and Changes in Nutritional Re-
quirements." *European Review for Medical and Pharmacological
Sciences* 17 (13): 1804–1813. https://pubmed.ncbi.nlm.nih
.gov/23852908.

26. Williams, A.-I., Cotter, A., Sabina, A., Girard, C., Goodman,
J., & Katz, D. L. (2005). "The Role for Vitamin B-6 as Treatment for Depression: A Systematic Review." *Family Practice* 22
(5): 532–537. https://doi.org/10.1093/fampra/cmi040.

27. Khalili, H., Granath, F., Smedby, K. E., Ekbom, A., Neovius, M., Chan, A. T., & Olen, O. (2016). "Association between Long-Term Oral Contraceptive Use and Risk of Crohn's Disease Complications in a Nationwide Study." *Gastroenterology* 150 (7): 1561–1567. https://doi.org/10.1053/j.gastro.2016.02.041.

28. Etminan, M., Delaney, J. A. C., Bressler, B., & Brophy, J. M. (2011). "Oral Contraceptives and the Risk of Gallbladder Disease: A Comparative Safety Study." *Canadian Medical Association Journal* 183 (8): 899–904. https://doi.org/10.1503/cmaj.110161.

29. Benagiano, G., Benagiano, M., Bianchi, P., D'Elios, M. M., & Brosens, I. (2019). "Contraception in Autoimmune Diseases." *Best Practice & Research Clinical Obstetrics & Gynaecology* 60: 111–123. https://doi.org/10.1016/j.bpobgyn.2019.05.003.

30. Williams, "Hormonal Contraception and the Development of Autoimmunity: A Review of the Literature."

31. Zimmerman, Y., Eijkemans, M. J., Coelingh Bennink, H. J., Blankenstein, M. A., & Fauser, B. C. (2014). "The Effect of Combined Oral Contraception on Testosterone Levels in Healthy Women: A Systematic Review and Meta-Analysis." *Human Reproduction Update* 20 (1): 76–105. https://doi.org/10.1093/humupd/dmt038.

32. Zimmerman et al., "The Effect of Combined Oral Contraception on Testosterone Levels in Healthy Women: A Systematic Review and Meta-Analysis."

33. Skovlund et al., "Association of Hormonal Contraception with Depression."

34. Barthelmess, E. K., & Naz, R. K. (2014). "Polycystic Ovary Syndrome: Current Status and Future Perspective." *Frontiers in Bioscience (Elite Edition)* 6 (1): 104–119. https://doi.org/10.2741/e695.

35. Jingjing Liu, Qunhong Wu, Yanhua Hao, Mingli Jiao, Xing Wang, Shengchao Jiang, & Liyuan Han. (2021). "Measuring the Global Disease Burden of Polycystic Ovary Syndrome in 194 Countries: Global Burden of Disease Study 2017." *Human Reproduction* 36 (4): 1108–1119. https://doi.org/10.1093/humrep/deaa371.

36. Barkley, G. S. (2008). "Factors Influencing Health Behaviors in the National Health and Nutritional Examination Survey,

III (NHANES III)." *Social Work in Health Care* 46 (4): 57–79. https://doi.org/10.1300/J010v46n04_04.

37. Franks, S., Gharani, N., Waterworth, D., Batty, S., White, D., Williamson, R., & McCarthy, M. (1997). "The Genetic Basis of Polycystic Ovary Syndrome." *Human Reproduction* 12 (12): 2641–2648. https://doi.org/10.1093/humrep/12.12.2641.

38. Kasim-Karakas, S. E., Cunningham, W. M., & Tsodikov, A. (2007). "Relation of Nutrients and Hormones in Polycystic Ovary Syndrome." *American Journal of Clinical Nutrition* 85 (3): 688–694. https://doi.org/10.1093/ajcn/85.3.688.

39. Basu, B. R., Chowdhury, O., & Saha, S. K. (2018). "Possible Link between Stress-Related Factors and Altered Body Composition in Women with Polycystic Ovarian Syndrome." *Journal of Human Reproductive Sciences* 11 (1): 10–18. https://doi.org/10.4103/jhrs.JHRS_78_17.

40. Dunaif, A. (1997). "Insulin Resistance and the Polycystic Ovary Syndrome: Mechanism and Implications for Pathogenesis." *Endocrine Reviews* 18 (6): 774–800. https://doi.org/10.1210/edrv.18.6.0318.

41. González, F. (2012). "Inflammation in Polycystic Ovary Syndrome: Underpinning of Insulin Resistance and Ovarian Dysfunction." *Steroids* 77 (4): 300–305. https://doi.org/10.1016/j.steroids.2011.12.003.

42. Gorpinchenko, I., Nikitin, O., Banyra, O., & Shulyak, A. (2014). "The Influence of Direct Mobile Phone Radiation on Sperm Quality." *Central European Journal of Urology* 67 (1): 65–71. https://doi.org/10.5173/ceju.2014.01.art4.

43. Chua, T.-E., Bautista, D. C., Tan, K. H., Yeo, G., & Chen, H. (2018). "Antenatal Anxiety: Prevalence and Patterns in a Routine Obstetric Population." *Annals of the Academy of Medicine, Singapore* 47 (10): 405–412. http://www.annals.edu.sg/pdf/47VolNo10Oct2018/MemberOnly/V47N10p405.pdf.

44. Linnakaari, R., Nelle, N., Mentula, M., Bloigu, A., Gissler, M., Heikinheimo, O., & Niinimäki, M. (2019). "Trends in the Incidence, Rate and Treatment of Miscarriage—Nationwide Register-Study in Finland, 1998–2016." *Human Reproduction* 34 (11): 2120–2128. https://doi.org/10.1093/humrep/dez211.

45. Declercq, E., & Zephyrin, L. (2021). "Maternal Mortality in the United States: A Primer." Commonwealth Fund, Decem-

ber 16. www.commonwealthfund.org/publications/issue-brief
-report/2020/dec/maternal-mortality-united-states-primer.

46. Centers for Disease Control and Prevention. (2021). "Work-
     ing Together to Reduce Black Maternal Mortality." Minority
     Health and Health Equity, CDC. www.cdc.gov/healthequity
     /features/maternal-mortality/index.html.

47. Berman, J. (2021). "Women's Unpaid Work Is the Backbone
     of the American Economy." Marketwatch, April 15. www
     .marketwatch.com/story/this-is-how-much-more-unpaid
     -work-women-do-than-men-2017-03-07.

48. Tolbert, J., Orgera, K., & Damico, A. (2020). "Key Facts about
     the Uninsured Population." KFF, November 6. https://www
     .kff.org/uninsured/issue-brief/key-facts-about-the-uninsured
     -population.

49. Mental Health America. (2021). "The State of Mental Health
     in America." www.mhanational.org/issues/state-mental-health
     -america.

50. Fairbrother, N., Janssen, P., Antony, M. M., Tucker, E., &
     Young, A. H. (2016). "Perinatal Anxiety Disorder Prevalence
     and Incidence." *Journal of Affective Disorders* 200: 148–155.
     https://doi.org/10.1016/j.jad.2015.12.082.

51. MGH Center for Women's Mental Health. (2015). "Is It
     Postpartum Depression or Postpartum Anxiety? What's the
     Difference?" September 30. https://womensmentalhealth.org
     /posts/is-it-postpartum-depression-or-postpartum-anxiety
     -whats-the-difference/.

52. Jamieson, D. J., Theiler, R. N., & Rasmussen, S. A. (2006).
     "Emerging Infections and Pregnancy." *Emerging Infectious
     Diseases* 12 (11): 1638–1643. https://pubmed.ncbi.nlm.nih.gov
     /17283611.

53. Khashan, A. S., Kenny, L. C., Laursen, T. M., Mahmood, U.,
     Mortensen, P. B., Henriksen, T. B., & O'Donoghue, K. (2011).
     "Pregnancy and the Risk of Autoimmune Disease." *PLoS ONE*
     6 (5). https://doi.org/10.1371/journal.pone.0019658.

Chapter 10: The Silent Epidemic

1.   Martin, C. B., Hales, C. M., Gu, Q., & Ogden, C. L. (2019).
     "Prescription Drug Use in the United States, 2015–2016."
     NCHS Data Brief No. 334, May. Centers for Disease

Control and Prevention. www.cdc.gov/nchs/products/data briefs/db334.htm.

2. "America's State of Mind Report." (2020). Express Scripts, April 16. https://www.express-scripts.com/corporate/americas -state-of-mind-report.

3. Christensen, J. C. (2021). "Benzodiazepines Might Be a 'Hidden Element' of the US Overdose Epidemic." CNN, January 20. www.cnn.com/2020/01/20/health/benzodiazepines -prescriptions-study/index.html.

4. Nemeroff, C. B. (2003). "The Role of GABA in the Pathophysiology and Treatment of Anxiety Disorders." *Psychopharmacology Bulletin* 37 (4): 133–146. https://pubmed.ncbi .nlm.nih.gov/15131523/.

5. Lydiard, R. B. (2003). "The Role of GABA in Anxiety Disorders." Supplement 3, *Journal of Clinical Psychiatry* 64: 21–27. https://pubmed.ncbi.nlm.nih.gov/12662130.

6. Griffin III, C. E., Kaye, A. M., Bueno, F. R., & Kaye, A. D. (2013). "Benzodiazepine Pharmacology and Central Nervous System–Mediated Effects." *Ochsner Journal* 13 (2): 214–223. https://www.ncbi.nlm.nih.gov/pmc/articles/PMC3684331.

7. Barnes Jr., E. M. (1996). "Use-Dependent Regulation of $GABA_A$ Receptors." *International Review of Neurobiology* 39: 53–76. https://doi.org/10.1016/s0074-7742(08)60663-7.

8. Higgitt, A., Fonagy, P., & Lader, M. (1988). "The Natural History of Tolerance to the Benzodiazepines." Monograph supplement, *Psychological Medicine* 13: 1–55. https://doi .org/10.1017/s0264180100000412.

9. Cookson, J. C. (1995). "Rebound Exacerbation of Anxiety during Prolonged Tranquilizer Ingestion." *Journal of the Royal Society of Medicine* 88 (9): 544. https://pubmed.ncbi.nlm.nih .gov/7562864.

10. Alexander, Scott. (2015). "Things That Sometimes Work if You Have Anxiety." *Slate Star Codex* (blog), June 13. https:// slatestarcodex.com/2015/07/13/things-that-sometimes-work -if-you-have-anxiety.

11. Davies, J., & Read, J. (2019). "A Systematic Review into the Incidence, Severity and Duration of Antidepressant Withdrawal Effects: Are Guidelines Evidence-Based?" *Addictive Behaviors* 97: 111–121. https://doi.org/10.1016/j.addbeh.2018.08.027.

12. Wilson, E., & Lader, M. (2015). "A Review of the Management of Antidepressant Discontinuation Symptoms." *Therapeutic Advances in Psychopharmacology* 5 (6): 357–368. https://doi.org/10.1177/2045125315612334.

Chapter 11: Discharging Stress and Cultivating Relaxation

1. Breit, S., Kupferberg, A., Rogler, G., & Hasler, G. (2018). "Vagus Nerve as Modulator of the Brain–Gut Axis in Psychiatric and Inflammatory Disorders." *Frontiers in Psychiatry* 9. https://doi.org/10.3389/fpsyt.2018.00044.
2. Tubbs, R. S., Rizk, E., Shoja, M. M., Loukas, M., Barbaro, N., & Spinner, R. J., eds. (2015). *Nerves and Nerve Injuries: Vol. 1: History, Embryology, Anatomy, Imaging, and Diagnostics* (Cambridge, MA: Academic Press).
3. Sengupta, P. (2012). "Health Impacts of Yoga and Pranayama: A State-of-the-Art Review." *International Journal of Preventive Medicine* 3 (7): 444–458. http://doi.org/10.13016/LXQD-LC0O.
4. Nemati, A. (2013). "The Effect of Pranayama on Test Anxiety and Test Performance." *International Journal of Yoga* 6 (1): 55–60. https://doi.org/10.4103/0973-6131.105947.
5. Roelofs, K. (2017). "Freeze for Action: Neurobiological Mechanisms in Animal and Human Freezing." *Philosophical Transactions of the Royal Society, Series B, Biological Sciences* 372. https://doi.org/10.1098/rstb.2016.0206.
6. Tsuji, H., Venditti Jr., F. J., Manders, E. S., Evans, J. C., Larson, M. G., Feldman, C. L., & Levy, D. (1994). "Reduced Heart Rate Variability and Mortality Risk in an Elderly Cohort. The Framingham Heart Study." *Circulation* 90 (2): 878–883. https://doi.org/10.1161/01.cir.90.2.878.
7. Buccelletti, E., Gilardi, E., Scaini, E., Galiuto, L., Persiani, R., Biondi, A., Basile, F., & Gentiloni Silveri, N. (2009). "Heart Rate Variability and Myocardial Infarction: Systematic Literature Review and Metanalysis." *European Review for Medical and Pharmacological Sciences* 13 (4): 299–307. https://pubmed.ncbi.nlm.nih.gov/19694345.
8. Taylor, S. E., Klein, L. C., Lewis, B. P., Gruenewald, T. L., Gurung, R. A. R., & Updegraff, J. A. (2000). "Biobehavioral Responses to Stress in Females: Tend-and-Befriend, Not

Fight-or-Flight." *Psychological Review* 107 (3): 411–429. https://doi.org/10.1037/0033-295x.107.3.411.

9. Taylor et al., "Biobehavioral Responses to Stress in Females," 412.

10. Taylor et al., "Biobehavioral Responses to Stress in Females," 413.

11. Kübler-Ross, Elisabeth, and Kessler, David. (2014). *On Grief and Grieving: Finding the Meaning of Grief through the Five Stages of Loss* (New York: Scribner), 66.

12. Konopacki, M., & Madison, G. (2018). "EEG Responses to Shamanic Drumming: Does the Suggestion of Trance State Moderate the Strength of Frequency Components?" *Journal of Sleep and Sleep Disorder Research* 1 (2): 16–25. https://doi.org/10.14302/issn.2574-4518.jsdr-17-1794.

13. Drisdale III, J. K., Thornhill, M. G., & Vieira, A. R. (2017). "Specific Central Nervous System Medications Are Associated with Temporomandibular Joint Symptoms." *International Journal of Dentistry*. https://doi.org/10.1155/2017/1026834.

14. Goodwin, A. K., Mueller, M., Shell, C. D., Ricaurte, G. A., & Ator, N. A. (2013). "Behavioral Effects and Pharmacokinetics of (±)-3,4-Methylenedioxymethamphetamine (MDMA, Ecstasy) after Intragastric Administration to Baboons." *Journal of Pharmacology and Experimental Therapeutics* 345 (3): 342–353. https://doi.org/10.1124/jpet.113.203729.

15. Fujita, Y., & Maki, K. (2018). "Association of Feeding Behavior with Jaw Bone Metabolism and Tongue Pressure." *Japanese Dental Science Review* 54 (4): 174–182. https://doi.org/10.1016/j.jdsr.2018.05.001.

16. De Moor, M. H., Beem, A. L., Stubbe, J. H., Boomsma, D. I., & De Geus, E. J. (2006). "Regular Exercise, Anxiety, Depression and Personality: A Population-Based Study." *Preventive Medicine* 42 (4): 273–279. https://doi.org/10.1016/j.ypmed.2005.12.002.

17. Byrne, A., & Byrne, D. G. (1993). "The Effect of Exercise on Depression, Anxiety and Other Mood States: A Review." *Journal of Psychosomatic Research* 37 (6): 565–574. https://doi.org/10.1016/0022-3999(93)90050-p.

18. Jayakody, K., Gunadasa, S., & Hosker, C. (2014). "Exercise for Anxiety Disorders: Systematic Review." *British Journal of*

Sports Medicine 48 (3): 187–196. https://pubmed.ncbi.nlm
.nih.gov/23299048.

19. Gleeson, M., Bishop, N., Stensel, D., Lindley, M. R., Mastana, S. S., & Nimmo, M. A. (2011). "The Anti-Inflammatory Effects of Exercise: Mechanisms and Implications for the Prevention and Treatment of Disease." *Nature Reviews Immunology* 11: 607–615. https://doi.org/10.1038/nri3041.

20. Jackson, E. (2013). "Stress Relief: The Role of Exercise in Stress Management." *ACSM's Health & Fitness Journal* 17 (3): 14–19. https://doi.org/10.1249/fit.0b013e31828cb1c9.

21. Harber, V. J., & Sutton, J. R. (1984). "Endorphins and Exercise." *Sports Medicine* 1 (2): 154–171. https://pubmed.ncbi .nlm.nih.gov/6091217.

22. McDonagh, B. (2015). *Dare: The New Way to End Anxiety and Stop Panic Attacks* (Williamsville, NY: BMD Publishing), 32.

23. McDonagh, *Dare*, 49.

Chapter 12: Tuning In

1. Brackett, Marc. (2019). *Permission to Feel: The Power of Emotional Intelligence to Achieve Well-Being and Success* (New York: Celadon Books), 11.

2. Miller, J. J., Fletcher, K., & Kabat-Zinn, J. (1995). "Three-Year Follow-Up and Clinical Implications of a Mindfulness Meditation–Based Stress Reduction Intervention in the Treatment of Anxiety Disorders." *General Hospital Psychiatry* 17 (3): 192–200. https://doi.org/10.1016/0163-8343(95)00025-m.

3. Hofmann, S. G., Sawyer, A. T., Witt, A. A., & Oh, D. (2010). "The Effect of Mindfulness-Based Therapy on Anxiety and Depression: A Meta-Analytic Review." *Journal of Consulting and Clinical Psychology* 78 (2): 169–183. https://doi.org /10.1037/a0018555.

4. Hofmann et al., "Effect of Mindfulness-Based Therapy."

5. Creswell, J. D., Way, B. M., Eisenberger, N. I., & Lieberman, M. D. (2007). "Neural Correlates of Dispositional Mindfulness during Affect Labeling." *Psychosomatic Medicine* 69 (6): 560–565. https://doi.org/10.1097/PSY.0b013e3180f6171f.

6. Singer, Michael. (2007). *The Untethered Soul: The Journey Beyond Yourself* (Oakland, CA: New Harbinger Publications), 10.

7. Tolle, Eckhart. (1999). *Practicing the Power of Now: Essential Teachings, Meditations, and Exercises from* The Power of Now (Novato, CA: New World Library), 40.

8. Kini, P., Wong, J., McInnis, S., Gabana, N., & Brown, J. W. (2016). "The Effects of Gratitude Expression on Neural Activity." *NeuroImage* 128: 1–10. https://doi.org/10.1016/j.neuroimage.2015.12.040.

9. Whitaker, Holly. (2019). *Quit Like a Woman: The Radical Choice Not to Drink in a Culture Obsessed with Alcohol* (New York: Dial Press), 151.

10. Moody, L., in conversation with Glennon Doyle. (2020). "Glennon Doyle on Overcoming Lyme Disease, Hope During Hard Times, and the Best Relationship Advice." *Healthier Together* (podcast), August 19. https://www.lizmoody.com/healthiertogetherpodcast-glennon-doyle.

11. Urban, M. (@melissau). (2021). "Six real-life boundaries I have recently set, word for word." Instagram, March 23. https://www.instagram.com/p/CMx0fWwsLmN.

12. Whitaker, *Quit Like a Woman*, 115.

13. Collignon, O., Girard, S., Gosselin, F., Saint-Amour, D., Lepore, F., & Lassonde, M. (2010). "Women Process Multisensory Emotion Expressions More Efficiently Than Men." *Neuropsychologia* 48 (1): 220–225. https://doi.org/10.1016/j.neuropsychologia.2009.09.007.

14. Marling, Brit. (2020). "I Don't Want to Be the Strong Female Lead." *New York Times Sunday Review*, February 7. https://www.nytimes.com/2020/02/07/opinion/sunday/brit-marling-women-movies.html.

15. Kübler-Ross, Elisabeth, and Kessler, David. (2014). *On Grief and Grieving: Finding the Meaning of Grief through the Five Stages of Loss* (New York: Scribner), 214.

16. Carhart-Harris, R. L., Leech, R., Hellyer, P. J., Shanahan, M., Feilding, A., Tagliazucchi, E., Chialvo, D. R., & Nutt, D. (2014). "The Entropic Brain: A Theory of Conscious States Informed by Neuroimaging Research with Psychedelic Drugs." *Frontiers in Human Neuroscience* 8: 20. https://doi.org/10.3389/fnhum.2014.00020.

17. Casey, Nell. (2012). "Just Don't Mention Timothy Leary." *Whole Living* (July–August), 66–71.

18. Fournier, J. C., DeRubeis, R. J., Hollon, S. D., Dimidjian, S., Amsterdam, J. D., Shelton, R. C., & Fawcett, J. (2010). "Antidepressant Drug Effects and Depression Severity: A Patient-Level Meta-Analysis." *JAMA* 303 (1): 47–53. https://doi.org/10.1001/jama.2009.1943. https://pubmed.ncbi.nlm.nih.gov/20051569/.

19. Goldberg, S. B., Pace, B. T., Nicholas, C. R., Raison, C. L., & Hutson, P. R. (2020). "The Experimental Effects of Psilocybin on Symptoms of Anxiety and Depression: A Meta-Analysis." *Psychiatry Research* 284. https://doi.org/10.1016/j.psychres.2020.112749.

20. Murrough, J., Iosifescu, D., Chang, L., Al Jurdi, R., Green, C., Perez, A., Iqbal, S., et al. (2013). "Antidepressant Efficacy of Ketamine in Treatment-Resistant Major Depression: A Two-Site Randomized Controlled Trial." *American Journal of Psychiatry* 170 (10): 1134–1142. https://doi.org/10.1176/appi.ajp.2013.13030392.

21. Mitchell, J. M., Bogenschutz, M., Lilienstein, A., Harrison, C., Kleiman, S., Parker-Guilbert, K., Ot'alora, M., et al. (2021). "MDMA-Assisted Therapy for Severe PTSD: A Randomized, Double-Blind, Placebo-Controlled Phase 3 Study." *Nature Medicine* 27: 1025–1033. https://doi.org/10.1038/s41591-021-01336-3.

22. Gasser, P., Kirchner, K., & Passie, T. (2014). "LSD-Assisted Psychotherapy for Anxiety Associated with a Life-Threatening Disease: A Qualitative Study of Acute and Sustained Subjective Effects." *Journal of Psychopharmacology* 29 (1): 57–68. https://doi.org/10.1177/0269881114555249.

23. Mash, D. C., Duque, L., Page, B., & Allen-Ferdinand, K. (2018). "Ibogaine Detoxification Transitions Opioid and Cocaine Abusers between Dependence and Abstinence: Clinical Observations and Treatment Outcomes." *Frontiers in Pharmacology* 9: 529. https://doi.org/10.3389/fphar.2018.00529.

24. Muttoni, S., Ardissino, M., & John, C. (2019). "Classical Psychedelics for the Treatment of Depression and Anxiety: A Systematic Review." *Journal of Affective Disorders* 258: 11–24. https://doi.org/10.1016/j.jad.2019.07.076.

25. Taylor, J., Landeros-Weisenberger, A., Coughlin, C., Mulqueen, J., Johnson, J. A., Gabriel, D., Reed, M. O., et al. (2017).

"Ketamine for Social Anxiety Disorder: A Randomized, Placebo-Controlled Crossover Trial." *Neuropsychopharmacology* 43 (2): 325–333. https://doi.org/10.1038/npp.2017.194.

26. Mitchell et al., "MDMA-Assisted Therapy."

27. Carhart-Harris, R., Giribaldi, B., Watts, R., Baker-Jones, M., Murphy-Beiner, A., Murphy, R., Martell, J., et al. (2021). "Trial of Psilocybin versus Escitalopram for Depression." *New England Journal of Medicine* 384 (15): 1402–1411. https://doi.org/10.1056/nejmoa2032994.

28. Spriggs, M. J., Kettner, H., & Carhart-Harris, R. L. (2021). "Positive Effects of Psychedelics on Depression and Wellbeing Scores in Individuals Reporting an Eating Disorder." *Eating and Weight Disorders* 26: 1265–1270. https://doi.org/10.1007/s40519-020-01000-8.

29. Brown, T., & Alper, K. (2017). "Treatment of Opioid Use Disorder with Ibogaine: Detoxification and Drug Use Outcomes." *American Journal of Drug and Alcohol Abuse* 44 (1): 24–36. https://doi.org/10.1080/00952990.2017.1320802.

30. Hart, Carl L. (2021). *Drug Use for Grown-Ups: Chasing Liberty in the Land of Fear* (New York: Penguin).

31. Criminal Justice Policy Organization. (2021). "Cannabis Policy (Marijuana)." www.cjpf.org/cannabis.

32. Dews, F. (2017). "Charts of the Week: Marijuana Use by Race, Islamist Rule in Middle East, Climate Adaptation Savings." Brookings, August 11. www.brookings.edu/blog/brookings-now/2017/08/11/charts-of-the-week-marijuana-use-by-race/.

33. Carhart-Harris, R. L., & Nutt, D. J. (2017). "Serotonin and Brain Function: A Tale of Two Receptors." *Journal of Psychopharmacology* 31 (9): 1091–1120. https://doi.org/10.1177/0269881117725915.

34. Carhart-Harris, R. L., Roseman, L., Bolstridge, M., et al. (2017). "Psilocybin for Treatment-Resistant Depression: fMRI-Measured Brain Mechanisms." *Science Reports* 7. https://doi.org/10.1038/s41598-017-13282-7.

35. Inserra, A., De Gregorio, D., & Gobbi, G. (2021). "Psychedelics in Psychiatry: Neuroplastic, Immunomodulatory, and Neurotransmitter Mechanisms." *Pharmacological Reviews* 73 (1): 202–277. https://doi.org/10.1124/pharmrev.120.000056.

36. Corne, R., & Mongeau, R. (2019). "Utilisation des psychédéliques en psychiatrie: Lien avec les Neurotrophines" [Neurotrophic Mechanisms of Psychedelic Therapy]. *Biologie Aujourd'hui* 213 (3–4): 121–129. https://doi.org/10.1051/jbio/2019015.

37. Flanagan, T. W., & Nichols, C. D. (2018). "Psychedelics as Anti-Inflammatory Agents." *International Review of Psychiatry* 30 (4): 363–375. https://doi.org/10.1080/09540261.2018.1481827.

38. Palhano-Fontes, F., Andrade, K. C., Tofoli, L. F., Santos, A. C., Crippa, J. A., Hallak, J. E., Ribeiro, S., & de Araujo, D. B. (2015). "The Psychedelic State Induced by Ayahuasca Modulates the Activity and Connectivity of the Default Mode Network." *PLoS ONE* 10 (2). https://doi.org/10.1371/journal.pone.0118143.

39. Siu, W. (@will.siu.md). (2020). "Psychedelics are much more than tools for healing trauma." Instagram, November 26. https://www.instagram.com/p/CID_MOZBtVm.

40. Ross, S., Bossis, A., Guss, J., Agin-Liebes, G., Malone, T., Cohen, B., Mennenga, S. E., et al. (2016). "Rapid and Sustained Symptom Reduction Following Psilocybin Treatment for Anxiety and Depression in Patients with Life-Threatening Cancer: A Randomized Controlled Trial." *Journal of Psychopharmacology* 30 (12): 1165–1180. https://doi.org/10.1177/0269881116675512.

41. Grob, C. S., Danforth, A. L., Chopra, G. S., Hagerty, M., McKay, C. R., Halberstadt, A. L., & Greer, G. R. (2011). "Pilot Study of Psilocybin Treatment for Anxiety in Patients with Advanced-Stage Cancer." *Archives of General Psychiatry* 68 (1): 71–78. https://doi.org/10.1001/archgenpsychiatry.2010.116.

42. Griffiths, R. R., Johnson, M. W., Carducci, M. A., Umbricht, A., Richards, W. A., Richards, B. D., Cosimano, M. P., & Klinedinst, M. A. (2016). "Psilocybin Produces Substantial and Sustained Decreases in Depression and Anxiety in Patients with Life-Threatening Cancer: A Randomized Double-Blind Trial." *Journal of Psychopharmacology* 30 (12): 1181–1197. https://doi.org/10.1177/0269881116675513.

43. Barrett, F. S., & Griffiths, R. R. (2018). "Classic Hallucinogens and Mystical Experiences: Phenomenology and Neural

Correlates." *Behavioral Neurobiology of Psychedelic Drugs* 36: 393–430. https://doi.org/10.1007/7854_2017_474.

44. Griffiths et al., "Psilocybin."

45. Davis, A. K., Barrett, F. S., May, D. G., Cosimano, M. P., Sepeda, N. D., Johnson, M. W., Finan, P. H., & Griffiths, R. R. (2020). "Effects of Psilocybin-Assisted Therapy on Major Depressive Disorder." *JAMA Psychiatry* 78 (5): 481–489. https://doi.org/10.1001/jamapsychiatry.2020.3285.

46. Belser, A., personal communication, August 2018.

Chapter 13: This Is Why You Stopped Singing

1. Eschner, K. (2021). "The Story of the Real Canary in the Coal Mine." *Smithsonian*, December 30. www.smithsonianmag .com/smart-news/story-real-canary-coal-mine-180961570/.

2. Chevalier, G., Sinatra, S. T., Oschman, J. L., Sokal, K., & Sokal, P. (2012). "Earthing: Health Implications of Reconnecting the Human Body to the Earth's Surface Electrons." *Journal of Environmental and Public Health*. https://doi.org/ 10.1155/2012/291541.

3. Wilson, Sarah. (2018). *First, We Make the Beast Beautiful: A New Journey through Anxiety* (New York: Dey Street), 165.

4. Thompson, D. (2021). "Workism Is Making Americans Miserable." *The Atlantic*, February 24. www.theatlantic.com/ideas /archive/2019/02/religion-workism-making-americans-miserable /583441/.

5. Moore, K. (2014). "Millennials Work for Purpose, Not Paycheck." *Forbes*, October 2. www.forbes.com/sites/karlmoore /2014/10/02/millennials-work-for-purpose-not-paycheck.

6. Vesty, L. (2016). "Millennials Want Purpose over Paychecks. So Why Can't We Find It at Work?" *The Guardian*, September 14. www.theguardian.com/sustainable-business/2016/sep /14/millennials-work-purpose-linkedin-survey.

7. Bertino, J. (2017). "Council Post: Five Things Millennial Workers Want More Than a Fat Paycheck." *Forbes*, October 26. www.forbes.com/sites/forbescoachescouncil/2017/10 /26/five-things-millennial-workers-want-more-than-a-fat -paycheck.

8. Thompson, "Workism."

9.   Wigert, B. (2020). "Employee Burnout: The Biggest Myth."
     Gallup, March 13. www.gallup.com/workplace/288539/employee
     -burnout-biggest-myth.aspx.

10. Brown, Brené. (@BreneBrown). (2015). "The danger of ex-
     haustion as a status symbol and productivity as a metric for
     self-worth." Twitter, March 4. https://twitter.com/BreneBrown
     /status/573209964119867392.

11. Klein, E., in conversation with Anne Helen Petersen and Derek
     Thompson. (2019). "Work as Identity, Burnout as Lifestyle."
     Vox Conversations (podcast), December 26. https://podcasts
     .apple.com/us/podcast/work-as-identity-burnout-as-lifestyle
     /id1081584611?i=1000436045971.

12. McKeown, G. (2014). *Essentialism: The Disciplined Pursuit of
     Less* (New York: Random House), 8.

13. Thompson, Derek. (2020). "How Civilization Broke Our
     Brains," review of *Work: A Deep History, from the Stone Age to
     the Age of Robots*, by James Suzman. *The Atlantic*, December
     13. https://www.theatlantic.com/magazine/archive/2021/01
     /james-suzman-work/617266.

Chapter 14: Connection Is Calming

1.   Shankar, A., Hamer, M., McMunn, A., & Steptoe, A. (2013).
     "Social Isolation and Loneliness." *Psychosomatic Medicine* 75
     (2): 161–170. https://doi.org/10.1097/psy.0b013e31827f09cd.

2.   Alcaraz, K. I., Eddens, K. S., Blase, J. L., Diver, W. R., Pa-
     tel, A. V., Teras, L. R., Stevens, V. L., et al. (2018). "Social
     Isolation and Mortality in US Black and White Men and
     Women." *American Journal of Epidemiology* 188 (1): 102–109.
     https://doi.org/10.1093/aje/kwy231.

3.   National Academies of Sciences, Engineering, and Medicine.
     (2020). "Risk and Protective Factors for Social Isolation and
     Loneliness," in *Social Isolation and Loneliness in Older Adults:
     Opportunities for the Health Care System* (Washington, DC:
     National Academies Press). www.ncbi.nlm.nih.gov/books
     /NBK557971/.

4.   Teo, A. R., Lerrigo, R., & Rogers, M. A. M. (2013). "The
     Role of Social Isolation in Social Anxiety Disorder: A
     Systematic Review and Meta-Analysis." *Journal of Anxiety*

Disorders 27 (4): 353–364. https://doi.org/10.1016/j.janxdis
.2013.03.010.

5. Venniro, M., Zhang, M., Caprioli, D., Hoots, J. K., Golden, S. A., Heins, C., Morales, M., Epstein, D. H., & Shaham, Y. (2018). "Volitional Social Interaction Prevents Drug Addiction in Rat Models." *Nature Neuroscience* 21 (11): 1520–1529. https://doi.org/10.1038/s41593-018-0246-6.

6. ScienceDaily. (2013). "Socially Isolated Rats Are More Vulnerable to Addiction, Report Researchers." January 23. www .sciencedaily.com/releases/2013/01/130123165040.htm.

7. Katie, B., & Mitchell, S. (2002). *Loving What Is: Four Questions That Can Change Your Life* (New York: Harmony Books), 2.

8. Roelofs, K. (2017). "Freeze for Action: Neurobiological Mechanisms in Animal and Human Freezing." *Philosophical Transactions of the Royal Society, Series B, Biological Sciences* 372. https://doi.org/10.1098/rstb.2016.0206.

9. Abdulbaghi, A., Larsson, B., & Sundelin-Wahlsten, V. (2007). "EMDR Treatment for Children with PTSD: Results of a Randomized Controlled Trial." *Nordic Journal of Psychiatry* 61 (5): 34–354. https://doi.org/10.1080/08039480701643464.

10. Marcus, S. V., Marquis, P., & Sakai, C. (1997). "Controlled Study of Treatment of PTSD Using EMDR in an HMO Setting." *Psychotherapy: Theory, Research, Practice, Training* 34 (3): 307–315. https://doi.org/10.1037/h0087791.

11. Rosenberg, Marshall. (2015). "Requesting That Which Would Enrich Life," chap. 6 in *Nonviolent Communication: A Language of Life*, 3rd ed. (Encinitas, CA: PuddleDancer Press).

12. Burdette, H. L., & Whitaker, R. C. (2005). "Resurrecting Free Play in Young Children: Looking beyond Fitness and Fatness to Attention, Affiliation, and Affect." *Archives of Pediatrics & Adolescent Medicine* 159 (1): 46–50. https://doi.org/10.1001 /archpedi.159.1.46.

13. Brown, S. L. (2014). "Consequences of Play Deprivation." *Scholarpedia* 9 (5): 30449. https://doi.org/10.4249/scholarpedia .30449.

14. Gray, P. (2011). "The Decline of Play and the Rise of Psychopathology in Children and Adolescents." *American Journal of Play* 3 (4): 443–463. https://www.psychologytoday.com /files/attachments/1195/ajp-decline-play-published.pdf.

15. Carmichael, M. S., Humbert, R., Dixen, J., Palmisano, G., Greenleaf, W., & Davidson, J. M. (1987). "Plasma Oxytocin Increases in the Human Sexual Response." *Journal of Clinical Endocrinology and Metabolism* 64 (1): 27–31. https://doi.org /10.1210/jcem-64-1-27.

16. Blum, Kenneth, Chen, Amanda L. C., Giordano, John, Borsten, Joan, Chen, Thomas J. H., Hauser, Mary, Simpatico, Thomas, Femino, John, Braverman, Eric R., & Barh, Debmalya. (2012). "The Addictive Brain: All Roads Lead to Dopamine." *Journal of Psychoactive Drugs* 44 (2): 134–143. https://doi.org/10.1080/02791072.2012.685407.

17. Antonelli, M., Barbieri, G., & Donelli, D. (2019). "Effects of Forest Bathing (Shinrin-Yoku) on Levels of Cortisol as a Stress Biomarker: A Systematic Review and Meta-Analysis." *International Journal of Biometeorology* 63 (8): 1117–1134. https://doi .org/10.1007/s00484-019-01717-x.

18. Li, Q. (2019). "Effets des forêts et des bains de forêt (shinrin-yoku) sur la santé humaine: Une revue de la littérature" [Effect of Forest Bathing (Shinrin-Yoku) on Human Health: A Review of the Literature]. *Santé publique* S1 (HS): 135–143. https://doi.org/10.3917/spub.190.0135.

19. Bratman, G., Hamilton, J., Hahn, K., Daily, G., & Gross, J. (2015). "Nature Experience Reduces Rumination and Subgenual Prefrontal Cortex Activation." *Proceedings of the National Academy of Sciences* 112 (28): 8567–8572. https://doi .org/10.1073/pnas.1510459112.

20. Berkowitz, R. L., Coplan, J. D., Reddy, D. P., & Gorman, J. M. (2007). "The Human Dimension: How the Prefrontal Cortex Modulates the Subcortical Fear Response." *Reviews in the Neurosciences* 18 (3–4): 191–207. https://doi.org/10.1515 /revneuro.2007.18.3-4.191.

21. Chevalier, G., Sinatra, S. T., Oschman, J. L., Sokal, K., & Sokal, P. (2012). "Earthing: Health Implications of Reconnecting the Human Body to the Earth's Surface Electrons." *Journal of Environmental and Public Health*. https://doi.org/10.1155 /2012/291541.

22. Kox, M., van Eijk, L. T., Zwaag, J., van den Wildenberg, J., Sweep, F. C., van der Hoeven, J. G., & Pickkers, P. (2014). "Voluntary Activation of the Sympathetic Nervous System and

Attenuation of the Innate Immune Response in Humans." *Proceedings of the National Academy of Sciences of the United States of America* 111 (20): 7379–7384. https://doi.org/10.1073/pnas.1322174111.

23. Mäkinen, T. M., Mäntysaari, M., Pääkkönen, T., Jokelainen, J., Palinkas, L. A., Hassi, J., Leppäluoto, J., et al. (2008). "Autonomic Nervous Function during Whole-Body Cold Exposure before and after Cold Acclimation." *Aviation, Space, and Environmental Medicine* 79 (9): 875–882. https://doi.org/10.3357/asem.2235.2008.

24. "Wim Hof Method." Accessed October 15, 2021. www.wimhofmethod.com.

25. Brown, Brené, interview with Barack Obama. (2020). "Brené with President Barack Obama on Leadership, Family and Service." In *Unlocking Us with Brené Brown* (podcast, 1:04),, December 7. https://brenebrown.com/podcast/brene-with-president-barack-obama-on-leadership-family-and-service.

Chapter 15: Holding On, Letting Go

1. Wilson, Sarah. (2018). *First, We Make the Beast Beautiful: A New Journey through Anxiety* (New York: Dey Street), 297.

2. Gilbert, Elizabeth. (2018). "I AM WILLING." Facebook, June 6. https://www.facebook.com/227291194019670/posts/i-am-willingdear-onesthis-picture-of-me-and-rayya-was-taken-one-year-ago-today-t/1850682221680551.

3. Oprah Winfrey, W. interview with Elizabeth Gilbert. (2019). "Elizabeth Gilbert Says: I Came Here to Live a Life, Fully, All of It | SuperSoul Sunday." OWN (YouTube video, 2:01), June 6. https://www.youtube.com/watch?v=q8E1gKuwS7I.

4. Brach, Tara. (2021). "A Heart That Is Ready for Anything." *Tara Brach* (blog), May 15. http://blog.tarabrach.com/2013/05/a-heart-that-is-ready-for-anything.html.

Appendix: Herbs and Supplements for Anxiety

1. Williams, A.-I., Cotter, A. Sabina, A., Girard, C., Goodman, J., & Katz, D. L. (2005). "The Role for Vitamin B-6 as Treat-

ment for Depression: A Systematic Review." *Family Practice* 22 (5): 532–537. https://doi.org/10.1093/fampra/cmi040.

2. Everett, J. M., Gunathilake, D., Dufficy, L., Roach, P., Thomas, J., Upton, D., & Naumovski, N. (2016). "Theanine Consumption, Stress and Anxiety in Human Clinical Trials: A Systematic Review." *Journal of Nutrition and Intermediary Metabolism* 4: 41–42. http://dx.doi.org/10.1016/j.jnim.2015.12.308.

3. Pratte, M. A., Nanavati, K. B., Young, V., & Morley, C. P. (2014). "An Alternative Treatment for Anxiety: A Systematic Review of Human Trial Results Reported for the Ayurvedic Herb Ashwagandha (*Withania somnifera*)." *Journal of Alternative and Complementary Medicine* 20 (12): 901–908. https://doi.org/10.1089/acm.2014.0177.

4. Mukai, T., Kishi, T., Matsuda, Y., & Iwata, N. (2014). "A Meta-Analysis of Inositol for Depression and Anxiety Disorders." *Human Psychopharmacology* 29 (1): 55–63. https://doi.org/10.1002/hup.2369.

5. Palatnik, A., Frolov, K., Fux, M., & Benjamin, J. (2001). "Double-Blind, Controlled, Crossover Trial of Inositol versus Fluvoxamine for the Treatment of Panic Disorder." *Journal of Clinical Psychopharmacology* 21 (3): 335–339. https://doi.org/10.1097/00004714-200106000-00014.

6. National Academies of Sciences, Engineering, and Medicine. (2017). "Mental Health," in *The Health Effects of Cannabis and Cannabinoids: The Current State of Evidence and Recommendations for Research* (Washington, DC: National Academies Press). www.ncbi.nlm.nih.gov/books/NBK425748/.

Index

About the Author

Ellen Vora, MD, is a holistic psychiatrist, acupuncturist, and yoga teacher. She takes a functional medicine approach to mental health—considering the whole person and addressing imbalance at the root. Dr. Vora received her BA from Yale University and her MD from Columbia University, and she is board-certified in psychiatry and integrative holistic medicine. She lives in New York City with her husband and daughter.